BARRIER CONTAINMENT TECHNOLOGIES FOR ENVIRONMENTAL REMEDIATION APPLICATIONS

(A project sponsored by E.I. DuPont de Nemours & Company)

Edited by

Ralph R. Rumer and Michael E. Ryan

A WILEY-INTERSCIENCE PUBLICATION

JOHN WILEY & SONS, INC.

New York • Chichester • Brisbane • Toronto • Singapore

The opinions, findings, and conclusions expressed are those of the author(s) and not necessarily those of the New York State Center for Hazardous Waste Management. Neither the Center nor any person acting on behalf of the Center assumes any liability with respect to the use of, or for damages resulting from the use of any information, apparatus, method or process disclosed in this book. The contents of this book are not to be used for advertising or promotional purposes. Citation of brand names does not constitute an official endorsement or approval of the use of such commercial products.

This text is printed on acid-free paper.

Library of Congress Cataloging in Publication Data:

Barrier containment technologies for environmental remediation applications principal contributors Andrew Bodocsi . . . [et al.]; edited by Ralph R. Rumer, Michael E. Ryan.

p. cm.

Includes bibliographical references and index.

ISBN 0-471-13272-1 (cloth : alk.paper)

1. In situ remediation. 2. Sealing (Technology) I. Bodocsi, A. II. Rumer, Ralph R. III. Ryan, Michael E.

TD192.8.B37 1995

628.5′2—dc20 95-30532

Printed in the United States of America

10 9 8 7 6 5 4 3 2 1

Principal Contributors

Andrew Bodocsi
University of Cincinnati
Cincinnati, OH

Jeffrey C. Evans
Bucknell University
Lewisburg, PA

David E. Daniel
University of Texas
Austin, TX

James K. Mitchell
Virginia Polytechnic Institute
and State University
Blacksburg, VA

DuPont Participants
(Containment and Transport Modeling Team)

Calvin C. Chien, Leader
Richard C. Landis
Terry D. Vandell

Prepared for

DuPont Corporate Remediation Group
300 Bellevue Parkway
Wilmington, Delaware

Prepared by

Ralph R. Rumer and Michael E. Ryan
New York State Center for Hazardous Waste Management
State University of New York at Buffalo
Buffalo, New York

CONTENTS

Preface **xi**

Abstract **xiii**

1 Introduction **1**

1.1 Remediation and Containment Technology 1
1.2 General Overview of Containment Technology Applications 3
1.3 Construction Quality Control 4

2 Overview of Contaminant Transport and Site Conditions **6**

2.1 Contaminant Transport Mechanisms 6
2.2 Geologic and Hydrogeologic Considerations 8
2.2.1 Stratigraphic and Hydrogeologic Considerations 8
2.2.2 Seismic Considerations 9
2.2.3 Groundwater Recharge and Discharge Zones 9

3 Vertical Barriers (Walls) **11**

3.1 Introduction 11
3.2 Applications of Vertical Barriers 11
3.3 Types of Vertical Barriers 14
3.3.1 Soil–Bentonite Slurry Trench Cutoff Walls 16
3.3.2 Cement–Bentonite Slurry Trench Cutoff Walls 18
3.3.3 Plastic Concrete Cutoff Walls 19
3.3.4 Diaphragm (Structural) Cutoff Walls 20
3.3.5 Vibrating Beam Cutoff Walls 21
3.3.6 Deep Soil Mixing 22
3.3.7 Composite Cutoff Walls 23
3.3.8 Steel Sheet Pile Cutoff Walls 24
3.3.9 Grouted Barriers 25
a. Permeation Grouting 25
b. Jet Grouting 30
c. Soilsaw™ Barrier Wall 31
3.4 Slurry Wall Performance Factors 32

3.4.1 Effect of Grain Size Distribution 32
3.4.2 Effect of Bentonite Content 33
3.4.3 Influence of Fluctuating Water Table 33
3.4.4 Effect of Consolidating Pressure 35
3.4.5 Field Stress Conditions 35
3.4.6 Hydraulic Fracturing Effects on Hydraulic Conductivity 40

3.5 Test Methods and Compatibility 41

3.5.1 Test Methods 41
3.5.2 Compatibility 42

3.6 Soil–Bentonite Backfill Design 48
3.7 Documentation of Performance 51
3.8 Construction Quality Control 51

3.8.1 Soil–Bentonite Cutoff Walls 51

a. Preconstruction 51
b. Construction 52

3.8.2 Cement–Bentonite Cutoff Walls 54

a. Preconstruction 55
b. Construction 55

3.8.3 Permeation Grouting 56

a. Preconstruction 56
b. Construction 56

3.8.4 Jet Grouting 57

a. Preconstruction 58
b. Construction 58

4 Bottom Barriers (Floors) 59

4.1 Introduction 59
4.2 Natural Bottom Barriers 60
4.3 Grouted Bottom Barriers 61

4.3.1 Grouts and Grouting 61
4.3.2 Construction of Bottom Barriers by Permeation Grouting 63
4.3.3 Construction of Bottom Barriers by Jet Grouting 65
4.3.4 Construction of Bottom Barriers by Directional Drilling and Grouting 67
4.3.5 Construction of Bottom Barriers Utilizing Hydrofracturing and Grouting 70
4.3.6 Construction of Bottom Barriers Using the Block Displacement Method 70

4.3.7 Sheet Pile Cutoff Combined with Injection Grouting 72
4.4 Other Bottom Barriers 74
4.4.1 The Zublin System 74
4.4.2 Microtunneling and Injection Grouting 76
4.4.3 Rectangular Microtubes Pushed Side by Side 77
4.5 Economic and Practical Considerations 78
4.6 Construction Quality Control 80

5 Surface Control Barriers (Caps) 83

5.1 Introduction 83
5.2 Components of a Final Cap 84
5.2.1 Surface Layer 84
5.2.2 Protection Layer 91
5.2.3 Drainage Layer 91
5.2.4 Barrier Layer 95
5.2.5 Foundation and/or Gas Collection Layer 110
5.3 Geosynthetic Clay Liners for Barrier Layer in Final Caps 110
5.3.1 Types of Geosynthetic Clay Liners 110
5.3.2 Engineering Properties 117
a. Hydraulic Conductivity 117
b. Shear Strength 122
c. Response to Differential Settlement 124
5.4 Interim Caps 126
5.5 Assessment for Overall Containment 126
5.5.1 Slope Angle 127
5.5.2 Annual Precipitation 128
5.5.3 Burrowing Animals 128
5.5.4 Components of a Cap 129
5.5.5 Compressibility of Underlying Waste 129
5.5.6 Control of Gas 129
5.6 Construction Quality Control 130
5.6.1 Compacted Soil 130
a. Preconstruction 130
b. Construction 132
5.6.2 Compacted Soil Augmented with Bentonite 134
a. Preconstruction 134
b. Construction 134
5.6.3 Geomembranes 134
a. Preconstruction 135

b. Construction 135
5.6.4 Geosynthetic Clay 136
a. Preconstruction 136
b. Construction 138
5.6.5 Postclosure Monitoring 138

6 Conclusions **139**

6.1 Barrier Material Formulations 140
6.2 Vertical Sidewall Barriers 141
6.3 Bottom Barriers 141
6.4 Surface Caps 142
6.5 Construction Quality Control 143
6.6 Long-Term Performance of Barrier Systems 143

References **144**

Glossary **156**

Index **163**

PREFACE

The activities of investigation and remediation of contaminated soil and groundwater (since the enactment of CERCLA, RCRA, and amendments) have led to the realization that, for certain contaminated sites, no technologically practical or economically affordable technologies or methods exist at the present time for accomplishing total remediation, contaminant removal, or destruction–degradation *in situ*. In these and perhaps other situations, *in situ* containment is an alternative that needs to be considered.

In recognition of this need, the Containment and Transport Modeling (C & TM) team, one of the five environmental remediation technology application teams established within the DuPont Corporate Remediation Group (CRG), conducted an extensive internal study culminating in a report, entitled, "Containment Technology Summary," issued in 1992.

Based on that earlier study, the C & TM team considered it necessary to further assess the state-of-the-art of containment technologies with the following specific objectives:

- Review and evaluate the current knowledge and practices of the containment technologies suitable for remediation.
- Identify areas where practical improvements could be developed, ensuring containment as a more reliable and acceptable alternative remediation measure.

Initiated by Calvin C. Chien, the C & TM team leader, the New York State Center for Hazardous Waste Management (NYSCHWM), located at the State University of New York at Buffalo, was contacted and Ralph R. Rumer, Professor of Civil Engineering and Executive Director of the Center, was requested to be Calvin Chien's counterpart in planning and carrying out the proposed review and evaluation study. Michael E. Ryan, Associate Professor of Chemical Engineering and Director of the Center's Business–Industry Affiliates Program, assisted throughout the entire study. Because of their expertise related to containment technologies, Andrew Bodocsi, David E. Daniel, Jeffrey C. Evans, and James K. Mitchell were invited by DuPont to participate in the study. Two members of the C & TM team, Terry D. Vandell and Richard C. Landis, were actively involved in the planning and review, and made technical contributions to the report. While all participants worked as a team and contributed to this report, initial drafts of the sections on vertical barriers, bottom barriers, and surface caps were prepared by Jeffrey C. Evans, Andrew Bodocsi, and David E. Daniel, respectively. James K. Mitchell was primarily responsible for reviewing sections of the report.

The initial planning for the study began in April 1992, the scope of the study was finalized in July of that year, and the first full project team meeting (including DuPont C & TM team members, NYSCHWM, and invited experts) was held in September 1992. The compilation, editing, and generation of this project report were the responsibility of Ralph R. Rumer, Michael E. Ryan, and Jane E. Warfield, all of NYSCHWM. The NYSCHWM would also like to acknowledge the invaluable help and assistance of Donald R. McMahon, David Millar, and Paul A. Witherspoon with various aspects of this project. After a number of meetings and review cycles, the final project report was completed in November 1993. That report, with slight modification, is presented here in book form.

The editors and contributors have dedicated their proceeds from the sale of this book to the following nonprofit institutions or organizations: E. I. DuPont de Nemours & Co. (to Nature Conservancy-Delaware); Ralph R. Rumer and Michael E. Ryan (to New York State Center for Hazardous Waste Management); Andrew Bodocsi (to University of Cincinnati); David E. Daniel (to University of Texas at Austin); Jeffrey C. Evans (to Bucknell University); and James K. Mitchell (to Virginia Polytechnic Institute and State University).

Buffalo, New York RALPH R. RUMER
MICHAEL E. RYAN
Cincinnati, Ohio ANDREW BODOCSI
Austin, Texas DAVID E. DANIEL
Lewisburg, Pennsylvania JEFFREY C. EVANS
Blacksburg, Virginia JAMES K. MITCHELL

ABSTRACT

The dearth of cost-effective remediation technologies and the current state-of-the-art associated with *in situ* remediation techniques prompted this investigation of containment technologies and strategies as a means for controlling contaminant source zones. The study was carried out in collaboration with the Containment and Transport Modeling Team, Dupont Corporate Remediation Group, Wilmington, DE, with the assistance of recognized geotechnical experts having experience in the field of containment technologies. The aim of the investigation was to provide a comprehensive review and evaluation of the state-of-the-art of containment technologies as presently practiced in remediation applications. The study was directed to providing an assessment of the advantages, limitations, and performance of each containment component, individually as well as in combination, and included the identification of research and development activities that would advance the state-of-the-art of containment technologies. The scope of the study was limited to the application of containment technologies at existing contaminated sites and did not address the design and construction of new waste disposal facilities.

This book is organized into three sections that examine the state of knowledge, constructability, and performance of the three main barrier types, that is, vertical barriers (walls), bottom barriers (floors), and surface barriers (caps). It is found that surface caps and vertical barrier walls can be engineered and constructed using present-day state-of-the-art knowledge, methods, and technologies. At sites where naturally occurring clay or rock formations can serve as effective bottom barriers, comprehensive containment becomes feasible by keying sidewall barriers into the bottom formation and into a constructed cap at the ground surface. The state-of-the-art with regard to construction of bottom barriers using engineered materials is not as developed as that for walls and caps, although several techniques have been demonstrated at the pilot scale while others are currently under development.

The final section of the book outlines specific research and development activities that would advance the state-of-the-art in the application of vertical walls, surface caps, and bottom floors in remediation applications. Additional observations are offered relating to construction quality control and to the prediction of the performance of barrier systems. The availability of a proven mathematical model for simulating transport and fate of contaminants through barriers incorporating diffusive and advective transport as well as retardation and degradation processes would greatly aid the design of effective containment systems and the specification of barrier formulations tailored to site-specific conditions.

Containment of a contaminant source zone may involve both active and passive control measures. Active control involves manipulation of hydraulic gradients to regulate the direction of groundwater flow while passive control involves the installation of physical barriers. Both limit the movement of the contaminant away from the source zone. Active and passive control measures can be employed concurrently. The most commonly employed types of barriers include slurry trench walls, grout barriers, and surface caps. Subsurface barriers have been used for decades in geotechnical applications to control groundwater seepage. Although slurry-wall barriers have been used in many cases as part of pump-and-treat remediation systems, they have not played a primary role in environmental remediation strategies.

The successful application of containment technologies requires knowledge of the hydrogeological conditions at the site and evaluation of the advantages, limitations, and costs associated with various containment options, giving careful attention to the compatibility of barrier materials with the contaminants present at the site. During construction of a containment system, quality control and assurance methods are required, as well as postconstruction monitoring to detect failures or flaws that might develop in the installed containment system. There is also a requirement for methods and technologies to manage and repair defects that may develop during the design lifetime of the containment system.

An effective barrier should reduce to acceptable levels the potential for both advective and diffusive transport of contaminants through the barrier. Containment barrier material formulations that act to retard the contaminants by sorption or other mechanisms offer the potential to further enhance the effectiveness of the barrier. Consideration has been given to constructing reactive containment barriers, presenting the possibility for biotic or abiotic degradation of contaminants that penetrate the barrier. The development of new and improved barrier formulations is a promising area for research and development.

The application of containment technologies offers a potentially cost-effective management option in situations where other alternative remediation options are presently cost-prohibitive or, although near the application stage, have not yet been proven reliable. There are other potentially important benefits that may be derived from successful containment applications, including the opportunity to consider other remediation options in parallel or at an appropriate future time when more cost-effective contaminant source degradation technologies are available. If necessary and appropriate, consideration can be given to using the contained subsurface system as an *in situ* reactor chamber, employing injection and extraction wells both to control hydraulic gradients and to supply needed nutrients and/or coreactants.

Improvements in the design, construction, and maintenance of containment systems will be forthcoming as a result of on-going research efforts aimed at the development of improved barrier wall formulations, placement technologies, and methods for ensuring quality in the constructed containment project. It is also likely that future applications of containment technologies will be carried out in combination with other innovative remediation technologies, thus offering new opportunities for achieving greater cost-effectiveness in remediation under site-specific situations.

BARRIER CONTAINMENT TECHNOLOGIES FOR ENVIRONMENTAL REMEDIATION APPLICATIONS

1 Introduction

1.1 REMEDIATION AND CONTAINMENT TECHNOLOGY

Three approaches for the prevention or control of subsurface contaminant migration that will immobilize the contaminant and/or render its impact to the environment to acceptable risk levels are (1) destruction or degradation of the contaminants by implementing some form of remediation technology, (2) stabilization of the contaminants, and (3) containment of the contaminants by the construction of a low-permeability barrier around the source zone combined with manipulation of hydraulic gradients.

Ideally, it would be desirable to develop innovative remediation technologies that are effective in remediating soil and groundwater quality to acceptable levels without being excessively complex or costly [U.S. Environmental Protection Agency (USEPA), 1989a]. The successful implementation of currently available remediation technologies is dependent on a variety of factors, including site specific conditions and the nature and amount of the particular contaminants that are present. Too often, remediation strategies have been unsuccessful, difficult to implement, or exceedingly costly. Consequently, the optimal use of limited available resources in accomplishing environmental remediation poses a formidable technical challenge. The dearth of cost-effective remediation technologies and the current state-of-the-art associated with *in situ* remediation or stabilization techniques have prompted renewed consideration of containment strategies as a means for controlling contaminant source zones.

The class of contaminants referred to as dense nonaqueous phase liquids (DNAPLs) present one of the most challenging remediation situations. The DNAPLs typically have low solubilities in water and high interfacial surface tensions. Because they are denser than water, they tend to move downward in the subsurface, ponding above low-permeability formations and becoming trapped in the pore system by surface tension forces. As a result, DNAPLs are both difficult to locate and to remove from the subsurface.

In some situations, containment may offer the needed flexibility in the allocation of limited resources and in the choice of alternative, and perhaps more cost-effective, remedial measures. Containment prevents contamination from becoming more widespread, as well as allowing for *in situ* treatment, and offers the potential for the application of promising cost-effective innovative technologies that are near the application stage, but not yet proven or available.

Containment barrier systems (i.e.,passive barriers to pollution migration)

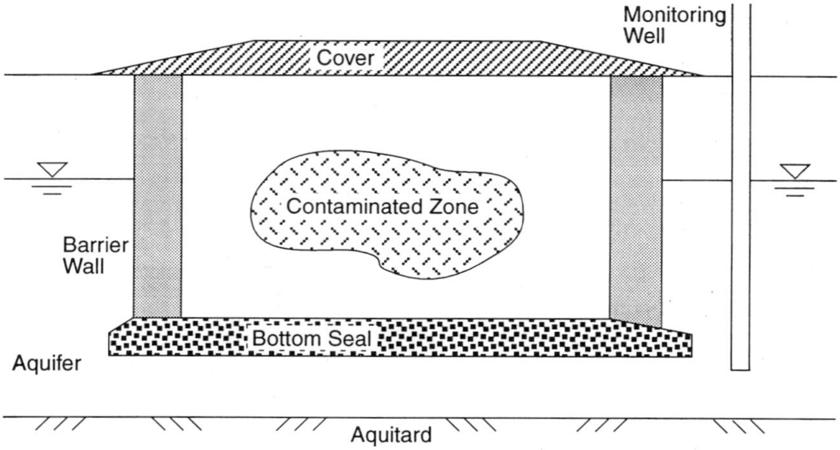

Figure 1.1 Total isolation using physical barriers for waste containment (after Mitchell and van Court, 1992b).

have been constructed to prevent or limit contaminant migration from both contaminated industrial sites as well as from newly constructed waste disposal facilities. The scope of this book is limited to the application of containment technologies at existing contaminated sites (e.g., leaking industrial landfills and buried drum sites). The design and construction of new waste disposal facilities is well-documented elsewhere (Daniel and Estornell, 1990; USEPA, 1987, 1989b, and 1989c).

A successful containment system should minimize, to acceptable risk levels, the potential hazards posed to human health and the environment. Contaminant migration control is accomplished by isolating the contaminant source zone utilizing low-permeability lateral and vertical migration control barriers. These physical containment systems could incorporate various functional components, including: (1) subsurface vertical barriers or walls to limit the lateral spread or migration of the contaminant; (2) subsurface bottom seals (floors) or vertical hydraulic control to limit the downward migration of a contaminant; and (3) an engineered surface cap or cover to minimize infiltration of surface water and to control the possible release or escape of volatile components (see Fig. 1.1). However, depending on the waste and site conditions, containment may only include the construction of a cap and/or side walls. The engineered cap also provides an effective barrier between the contaminated site and the surface. Each of these components is considered in detail in subsequent sections of this book.

Although subsurface barriers can be constructed to have very low hydraulic conductivities, they do not constitute totally impermeable systems. Hydraulic or active control measures (e.g., extraction and/or injection wells) may be used in conjunction with one or more passive barrier components to create inward hydraulic gradients that further minimize the potential for lateral and vertical contaminant migration. The ingredients of the material used to construct the barrier may also inhibit contaminant transport by chemically or physically binding the contaminant within the barrier material matrix (Cadena, 1989; Bierck and Chang, 1994; Mott and Weber, 1992).

The aim of this book is to provide a comprehensive review and evaluation of the state-of-the-art of containment technologies as presently practiced. This book is directed at providing an assessment of the advantages, limitations, and performance of each containment component, individually as well as in combination, in order to provide a basis for identifying areas where practical improvements can be developed, ensuring containment as a more reliable and acceptable alternative remediation measure.

The information contained in this book may assist environmental remediation professionals in considering containment technology as a practical and economical permanent solution or as a complementary interim measure. An objective and comprehensive evaluation of the application and performance of containment technologies is an essential step in developing a basis for future implementation and wider acceptance of this remediation strategy.

1.2 GENERAL OVERVIEW OF CONTAINMENT TECHNOLOGY APPLICATIONS

Containment of a contaminant source zone may involve both active and passive control measures. Active control involves manipulation of hydraulic gradients to control the direction of groundwater flow while passive control involves the installation of physical barriers. Both limit the movement of the contaminant away from the source zone. Active and passive control measures can be employed concurrently. The most commonly employed types of barriers include slurry trench walls, grout barriers, and surface caps. Subsurface barriers have been used for decades in the construction industry in order to reduce groundwater seepage under, through, or into underground structures or excavations (Xanthakos, 1994). Traditionally, soil–bentonite slurry mixes and grouts have been used as seepage barriers in the construction of dams and as sealants in the construction of rock tunnels, respectively.

More recently, barrier systems have been employed for contaminant migration control or as an integral component of a hazardous waste site remediation effort (Spooner et al., 1984). Many Records of Decision (RODs) involving Superfund sites have specified a vertical barrier or cutoff wall as a component of the overall site remediation plan. Barrier walls can be placed upgradient from a contaminated site to prevent flow of uncontaminated groundwater into the contaminated region, downgradient to prevent offsite flow of contaminated water, or around a source to contain the contaminants and to reduce the quantity of groundwater to be treated by inhibiting the inflow of uncontaminated groundwater. A barrier wall generally needs to be extended downward and keyed into a formation having a low hydraulic conductivity; however, hydraulic control can be implemented, and in situations where the contaminant is lighter than water [e.g., light nonaqueous phase liquids (LNAPLs)], the barrier wall may need to extend only partly below the water table to prevent migration.

At some contaminated sites, where an underlying low permeability natural barrier exists at reasonable depth below the surface, the most cost-effective option may be to key the vertical barrier into the natural underlying low permeability strata. When this is not possible, consideration may need to be given to the placement of a bottom barrier. However, the design and construction of bottom barriers (or floors) for containment purposes has been limited.

The successful application of containment technologies requires knowledge of the hydrogeological conditions at the site; assessment of the advantages, limitations, and costs associated with various containment technology options; and evaluation of the compatibility of barrier materials with the contaminants present at the site. In addition, the following are required: (1) methods for quality control and assurance during construction of the containment system; (2) installation of an appropriate monitoring system for detecting failures of the installed containment system; and (3) technologies for management and repair of defects that may develop during the design lifetime of the containment system. The emplacement of gelling liquids may be effective in sealing leaks in barrier walls (Persoff et al., 1994).

1.3 CONSTRUCTION QUALITY CONTROL

Choices must be made during design regarding selection of the type of barrier to contain waste deposits or contaminants. Construction quality control (CQC) affects these decisions since different barrier types utilize different methods for testing the quality of the barrier, each providing a different level of confidence. Some CQC methods are more effective than others, some are more costly, and some provide a higher degree of confidence in the quality of the constructed project.

The term CQC is often used interchangeably with construction quality assurance (CQA) and the two are frequently confused. The United States Environmental Protection Agency (USEPA, 1987) defines quality control as the contractor's observations, sampling, measuring, and testing to establish conformance to plans and specifications. Quality assurance, which is usually conducted by or for the project engineer, provides additional assurance to the owner that the quality of construction conforms to plans and specifications. Specific activities and frequencies for quality assurance functions are not mandated (USEPA, 1987).

The term CQC will be used to refer to both the CQC and CQA activities and functions employed to measure quality in the constructed project. The

CQC activities are summarized for the different types of barriers in subsequent chapters of the book.

There are two common concerns in CQC programs for all three forms of passive containment barriers (walls, floors, and caps). First, the material used to construct the physical barrier must exhibit an in-place hydraulic conductivity less than or equal to that given in the design specifications. In this regard, CQC requires testing the materials for permeability; however, this testing is often done in a laboratory using small samples of the selected material. The second concern relates to the continuity of the physical barrier (Grube, 1992a). Testing to ensure continuity varies among the different forms of barriers.

One of the design steps involves preparation of a written CQC plan and a CQA plan (Goldman et al., 1990) which are developed to identify the test methods and observations that will be used during construction to detect the occurrence of potential failure modes. The CQC plan specifies the reporting structure, CQC personnel qualifications, sampling and testing methods, required observations, testing frequency, acceptance criteria, corrective measures, and documentation requirements. Each CQC plan should be developed for the specific construction method used.

Tests included in a CQC program must provide the information necessary for field inspectors to make timely decisions regarding the acceptability of the barrier. Testing techniques have been introduced that may provide accurate measures of the constructed barrier's properties but cannot provide the data in a timely manner to enable field inspectors to use the results for acceptability decisions. For example, infiltrometer tests have been developed to measure the in-place permeability of compacted soil. However, these tests take several weeks to perform (Daniel, 1989). Thus, these testing techniques have limited application in practice. Further discussion of applicable testing techniques is presented in subsequent chapters.

2 Overview of Contaminant Transport and Site Considerations

2.1 CONTAMINANT TRANSPORT MECHANISMS

In the unsaturated zone, contaminant transport is influenced primarily by the effective porosity, permeability, texture, moisture content, soil chemistry, and various physical, chemical, and biological characteristics of the contaminants. Volatile components are transported through the open pore space by means of vapor phase flow. Liquids percolate through the pore space under the influence of gravity and capillary forces. Precipitation, infiltration, and downward percolation of water through the unsaturated zone will enhance the transport of contaminants toward the water table or capillary fringe. Precipitation, which infiltrates into a waste site, and any liquids already present combine to form the mixture that is commonly referred to as leachate.

Under saturated conditions, dissolved contaminants are transported mainly by advective movement of the bulk fluid resulting from the hydraulic gradients present in the saturated zone. Advective transport is always accompanied by molecular diffusion wherever concentration gradients are present and by mechanical dispersion that results from the tortuosity of the pore system geometry.

In the absence of significant advection (i.e., when hydraulic gradients are very small or absent), transport of dissolved contaminants still occurs if concentration gradients are present (see Fig. 2.1). However, for most saturated aquifers, advective transport is generally much larger than diffusive transport. Transport of dissolved ions can also occur under the action of electrical and thermal potential gradients; however, these are generally not important transport mechanisms under natural conditions.

Advective–diffusive transport may be affected by sorption processes that act to partition the dissolved contaminant (solute) between the liquid, gaseous, and solid phases present. Biochemical reactions may also occur, causing precipitation and/or transformations in the chemical form of the contaminant.

An effective containment barrier should minimize to acceptable levels the potential for contaminant transport by the above mechanisms. Since advective transport is generally the principal means for contaminant transport, containment barriers need to have extremely low hydraulic conductivities, less than 10^{-7} cm/s. This level is generally considered adequate to prevent significant

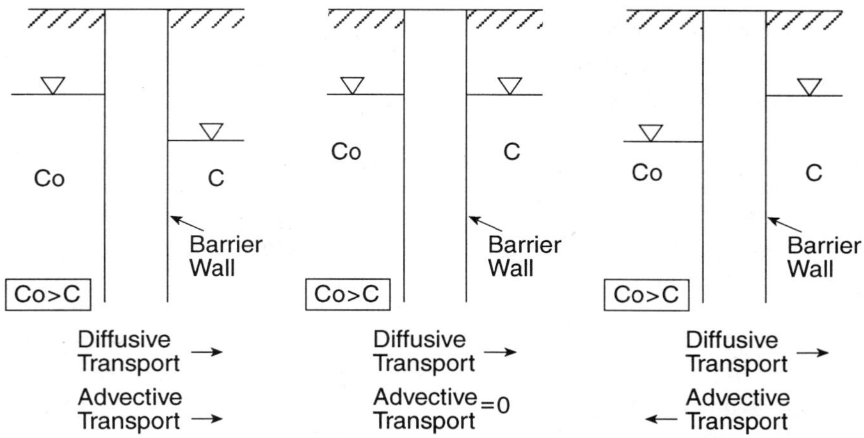

Figure 2.1 Scenarios for advective and diffusive transport at a sidewall barrier (after Manassero and Pasqualini, 1993).

chemical transport by advection, and can be realistically achieved with currently available materials and design practices. Equally important, the containment barrier must be constructed with materials that will not be adversely affected by the chemistry of the contaminant, causing possible increases in the hydraulic conductivity with time.

Even in the absence of advective transport caused by hydraulic gradients, Mott and Weber (1991a and b) have shown that molecular diffusion can result in significant migration of low molecular weight solutes in soil–bentonite barriers. They suggested that the effective diffusivities of solutes in soil–bentonite barriers could be estimated through modification of the corresponding free aqueous diffusivity based on knowledge of the barrier porosity. Mott and Weber (1992) investigated the reduction of diffusive transport that resulted from the addition of fly ash to the soil–bentonite barrier and found that the residual carbon typically associated with some fly ash from older power plants acted as an effective sorbent. Their results indicate that the addition of sorbents to barrier formulations can effectively retard diffusive transport, thus greatly extending the effective design life of a containment barrier wall.

Containment barrier materials that display sorption or other binding mechanisms (retardation) offer the potential to further reduce contaminant transport. Soil–bentonite mixtures augmented with granular activated carbon (GAC) have been found to be more effective in retarding transport of an organic solute (2-pentanone) than mixtures not containing GAC (Bierk and Chang, 1994). Some consideration has been given to constructing reactive containment barriers, thus offering the potential for biotic or abiotic degradation of contaminants that penetrate a barrier (Krumholz, 1993; Stokes, 1993; Gillham and O'Hannesin, 1992; O'Hannesin and Gillham, 1992; Gillham et al., 1993; Starr and Cherry, 1994).

2.2 GEOLOGIC AND HYDROGEOLOGIC CONSIDERATIONS

2.2.1 Stratigraphic and Hydrogeologic Considerations

Stratigraphic and hydrogeologic features of subsurface formations, and associated heterogeneities within and adjacent to a contaminant source zone, must be well defined prior to designing a containment system (Delage and Mazeas, 1993). Detailed delineation of the subsurface is needed to ensure proper and effective placement of containment barrier walls and floors.

The presence of boulders or other subsurface obstructions may make the excavation of slurry trenches (or other subsurface excavation for containment) difficult or impossible. If containment is to be accomplished using permeation grouting methods, the pore size of the material to be grouted must be considered. Generally, fine-grained soils are not suitable for permeation grouting, since the grout cannot readily penetrate the very small interparticle voids.

Factors that determine the rate at which a contaminant plume migrates through the subsurface also need to be considered in barrier wall design (Starr and Cherry, 1990). As mentioned previously, these factors include permeability and effective porosity, hydraulic gradient, interactions between the dissolved materials and the soil, and chemical or biological transformation reactions. The particular reactions that will occur within the containment barrier material need to be identified.

Vertical and lateral stratigraphic and structural discontinuities (e.g., regions of higher permeability, particularly extensive or continuous faults, fracture, or solution channel zones) can have a significant impact on containment wall and floor barrier material placement techniques, installation costs, and barrier integrity. Highly permeable zones can, during construction, channel pressurized grout away from the original design boundaries of the containment area, making construction of the containment wall difficult and expensive. If a continuous grout seal cannot be achieved, the resulting containment structure may be ineffective in controlling contaminant migration and various other options may need to be considered. In such cases, for example, vertical sheet piling may need to be placed, followed by a grout or bentonite slurry, to help minimize leakage through the sheet pile interlocks.

Lithologic and stratigraphic features can, however, also be used to advantage. The presence of a naturally occurring laterally continuous geologic unit(s) of suitable thickness with low hydraulic conductivity (i.e., $< 10^{-7}$ cm/s), underlying a contaminant source zone, may serve very effectively as a containment floor. Significant cost savings can sometimes be realized by tying vertical containment walls into such naturally occurring low-permeability continuous geologic units.

Even when a zone of low hydraulic conductivity is not located immediately below a source zone, but is located within an acceptable (cost effective) depth below the contaminant source area, it may be feasible to use this deeper formation as the containment floor. This can be accomplished by extending the vertical containment barrier wall below the vertical extent of significant contamination (source zone), through the lesser contaminated zone, into the underlying low permeable zone. This approach may not be feasible if the intermediate zone with reduced contaminant levels located immediately below the source zone, but above the natural formation (containment floor), is environmentally sensitive (e.g., is part of a significant aquifer flow system).

Active hydraulic control (pumping from within the containment boundary area) is also sometimes recommended to further prevent contaminant migration. Inward hydraulic gradients prevent advective transport of the contaminant through the containment barrier and minimize contaminant contact with the barrier material, thus reducing the potential for degradation of the barrier wall material.

2.2.2 Seismic Considerations

At waste containment sites located in seismically active regions, the potential for release of contaminants (particularly hazardous or toxic waste material) due to possible seismic activity should always be evaluated. Detailed geologic and structural features (faults, fractures) need to be well defined in the containment area and surrounding region. A detailed review of hazardous waste facilities in seismic zones has been given by Inyang (1991).

Regional seismic records and related publications can generally be obtained from the U. S. Geological Survey, state regulatory agencies, or from nearby universities. Available seismic information should be evaluated early in the containment design phase.

Contaminated sites located in areas of high seismic activity may not be suitable for containment technology applications. Source removal or *in situ* treatment may be necessary in these areas. In some situations, diaphragm walls can be used in areas where high structural stability is required (Mitchell, 1986). These walls are constructed in a manner similar to slurry trenches, where the clay slurry is replaced with precast or cast-in-place concrete panels.

2.2.3 Groundwater Recharge and Discharge Zones

The presence of major groundwater recharge or discharge zones needs to be considered in the design of a containment facility. These zones can present potential problems during the construction and operational phases of a containment facility.

Major groundwater recharge zones or discharge zones (e.g., near wetlands, springs, and pumping wells) frequently increase the local hydraulic gradients and, as a result, the groundwater flow velocities at a site. Steep hydraulic gradients and high groundwater flow rates can pose problems during the containment construction phase. Difficulty in grout placement due to grout migration, and potentially poor grout setup, could impact containment wall construction and integrity. Hydraulic control by aquifer pumping can help to reduce adverse hydraulic effects during construction, but with an increase in project construction costs.

As discussed by Mitchell (1986), regional and local groundwater flow regimes must be considered in selecting the type and location of barrier walls. Barrier walls may be placed upgradient, downgradient, or circumferentially around the contaminated zone, depending on the remediation objective.

3 Vertical Barriers (Walls)

3.1 INTRODUCTION

In this chapter the state of knowledge regarding vertical barriers is examined, areas of certainty and uncertainty are identified, and aspects requiring further study are suggested. This examination focuses on current vertical barrier technologies (such as soil–bentonite and cement–bentonite) but also includes potential new technologies not widely employed at present (such as composite cutoff walls).

Design and construction information that is considered ''common knowledge'' and is widely available elsewhere is not repeated here. For this background, the reader is referred to several existing publications on the subject (Boyes, 1975; Evans 1991 and 1993; LaGrega et al., 1994; Manassero and Pasqualini, 1993; Millet and Perez, 1981; Ryan, 1987; Spooner et al., 1984). However, for the sake of completeness and to permit discussion of design, construction, and performance parameters, a brief summary is provided of the various types and applications of vertical barriers.

3.2 APPLICATIONS OF VERTICAL BARRIERS

Vertical barriers function in the subsurface to contain contaminants and/or to redirect groundwater flow (seepage barrier). A vertical barrier wall around a site diminishes the rate of contaminant migration from the site, particularly in anisotropic materials having a high horizontal hydraulic conductivity. Passive barriers have been widely used for containment abroad (Jessberger, 1993a; Manassero and Pasqualini, 1993) but have not been routinely employed as a principal method of hazardous waste site remediation in the United States. However, once reduction in contaminant levels at sites undergoing cleanup by ''pump and treatment'' methods has reached acceptable levels, a vertical barrier wall used as a system component and left in place will, in the long term, serve to reduce the rate at which any residual contaminants migrate.

A common configuration for vertical barriers is as a circumferential cutoff wall where the vertical barrier completely surrounds the waste or the waste site as shown in Fig. 3.1. An important function of the vertical barrier in a remediation system, besides containment, is to restrict the flow of uncontaminated (or clean) groundwater into the site. In performing this function, vertical bar-

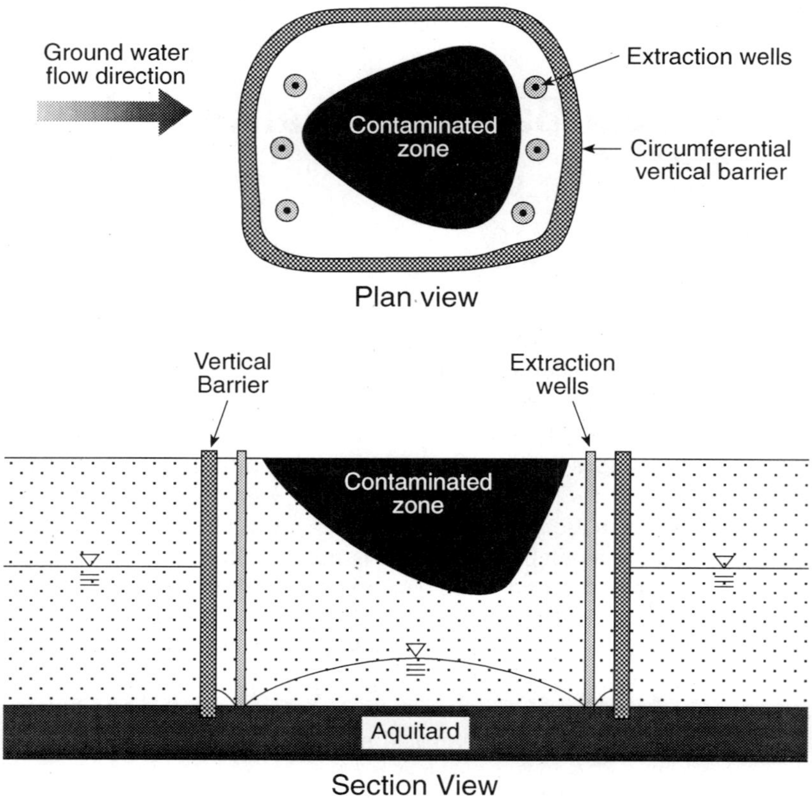

Figure 3.1 Circumferential vertical barrier.

riers are commonly used in conjunction with groundwater extraction and treatment systems. This is illustrated in Fig. 3.1 in which groundwater is extracted from within the cutoff wall and sent to treatment. Without the installation of a vertical barrier, clean groundwater would flow into or beneath the site, become contaminated, and increase the volume of water to be pumped and treated. Vertical barriers also provide groundwater control during construction operations for direct treatment of waste, waste removal, or placement of liner systems. Once construction operations have been completed, the barrier left in place will provide secondary protection against contaminant migration.

An upgradient vertical barrier wall (Fig. 3.2) provides a cutoff for controlling the influx of clean groundwater from upgradient regions. In conjunction with a pumping system located on the downgradient side (and no vertical barrier on the downgradient side of the site), groundwater withdrawal may recover contaminants that have migrated downgradient of the site. In this way, contaminants that have left the site, may be captured. A downgradient cutoff wall,

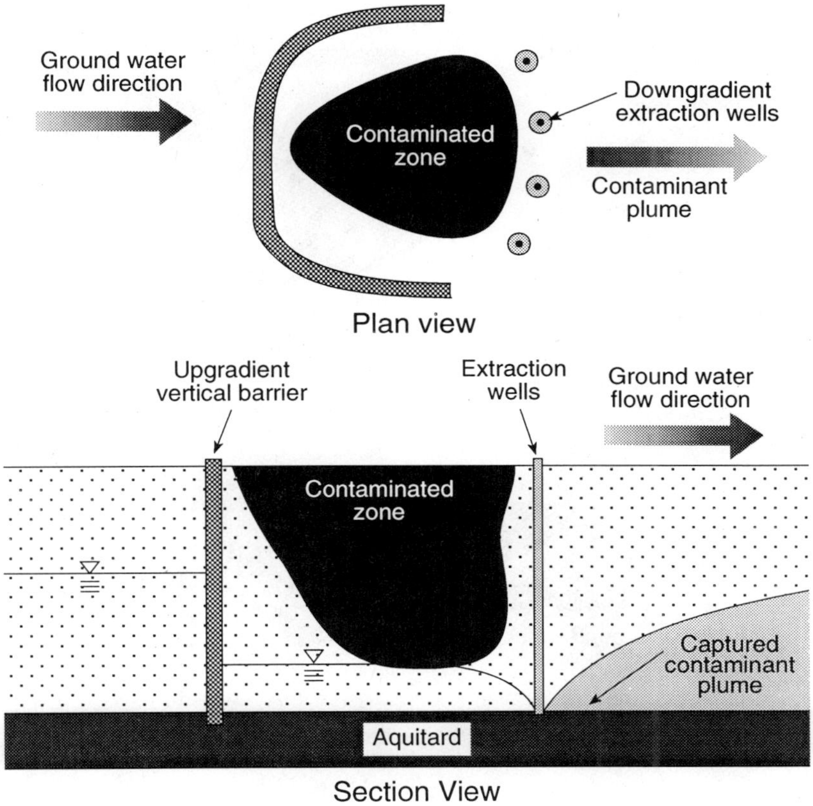

Figure 3.2 Upgradient vertical barrier.

as shown in Fig. 3.3, allows groundwater flowing through the site to flush contaminants from the site.

Vertical barriers are usually, but not always, embedded, or keyed into a material of low permeability beneath the site (see Fig. 3.4a). For LNAPLs, such as oil product leaked from a fuel oil storage tank, the vertical barrier does not need to fully penetrate the aquifer (see Fig. 3.4b). Such walls are termed "hanging walls".

A level of hydraulic conductivity equal to or less than 10^{-7} cm/s is normally specified for soil–bentonite slurry trench cutoff walls, one of the most common types of vertical barriers, since this is often adequate to prevent significant chemical transport by advection and is a level that can be readily achieved with careful design and construction. Calculations of site specific permissible flow rates or contaminant transport rates may provide a rational guide to the selection of the characteristics and type of vertical barrier to be used. As an ex-

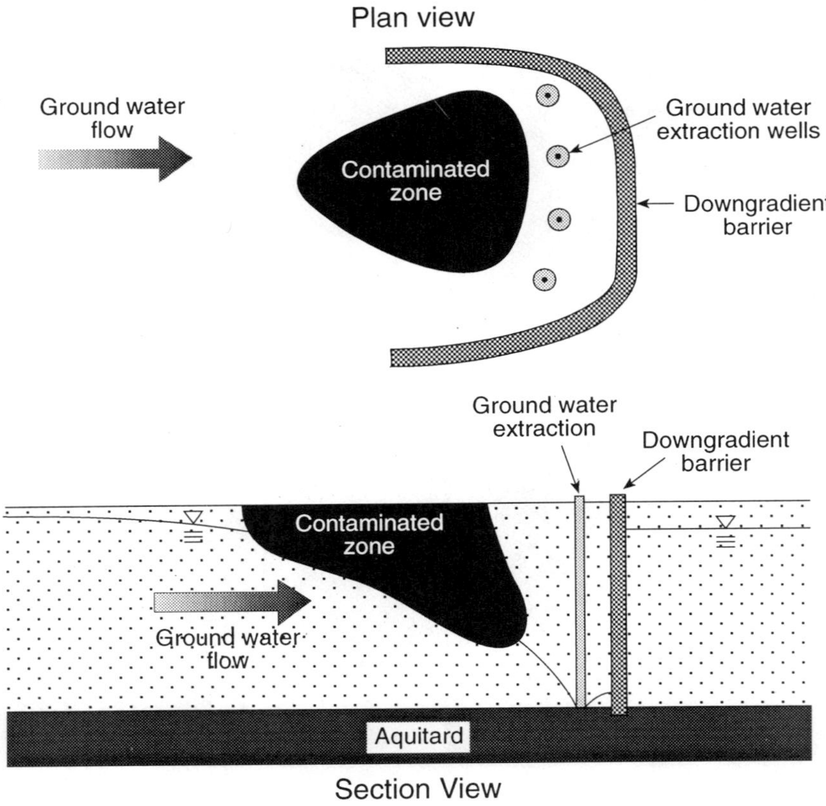

Figure 3.3 Downgradient vertical barrier.

ample, consider the construction of a 3-ft thick circumferential barrier to control the influx of clean groundwater to extraction wells (typical pump and treat applications). If the optimum, cost effective treatment plant capacity is 5 gpm, then for a barrier with a perimeter of 3000 ft, aquifer thickness of 30 ft, inward gradient of 1.67 across the barrier wall of 3-ft thickness (established by lowering the water level within the containment area by ~ 5 ft), the barrier must have a hydraulic conductivity of 2.3×10^{-6} cm/s. This suggests that when barrier walls are constructed to reduce inward flow of uncontaminated water, the normal requirement that the hydraulic conductivity be equal to or less than 10^{-7} cm/s may be relaxed.

3.3 TYPES OF VERTICAL BARRIERS

The barrier wall function and configuration are coupled with site and subsurface conditions in the selection of the suitable type(s) of vertical barrier. For

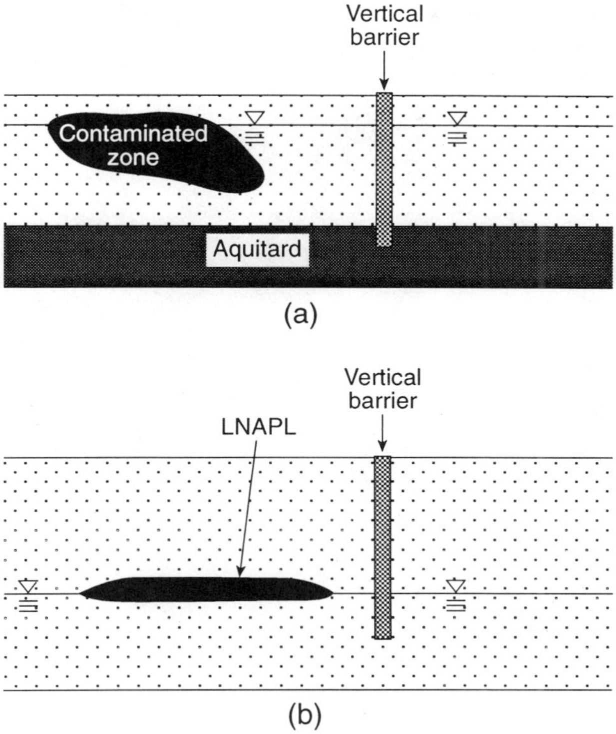

Figure 3.4 Embedment of vertical barrier walls. (a) Keyed vertical barrier wall and (b) hanging vertical barrier wall.

example, grout curtains are appropriate for vertical barriers in fractured rock, whereas, slurry trench cutoff wall techniques typically require softer materials that lend themselves to soil excavation technologies. Soil–bentonite slurry trench cutoff walls are the most common wall type used for Superfund site remediation (Ward, 1984). Each of the following techniques is discussed in more detail in this section:

- Soil–bentonite slurry trench cutoff walls.
- Cement–bentonite slurry trench cutoff walls.
- Plastic concrete cutoff walls.
- Diaphragm cutoff walls.
- Vibrating beam cutoff walls.
- Deep soil-mixed walls.
- Composite cutoff walls.
- Steel sheet pile cutoff walls.
- Grouted barriers.

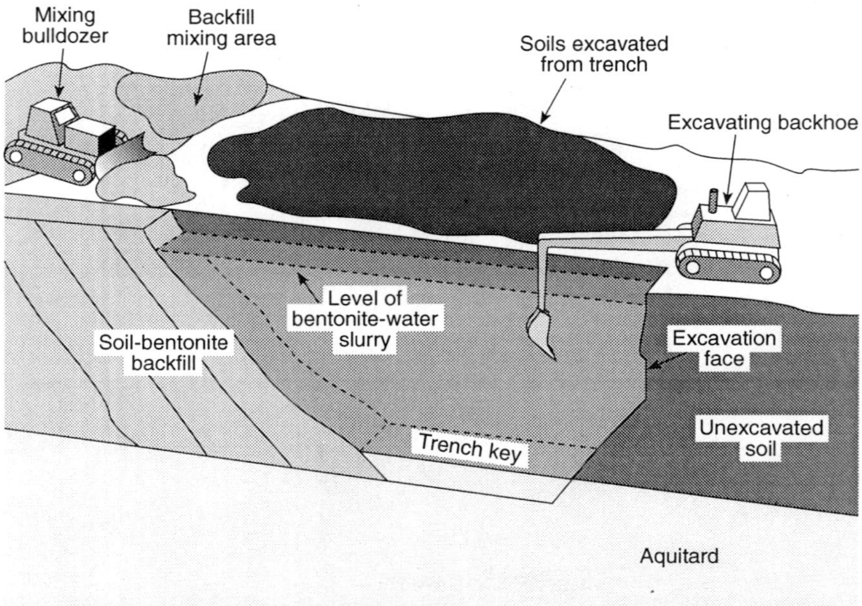

Figure 3.5 Soil–bentonite slurry trench cutoff wall construction.

3.3.1 Soil–Bentonite Slurry Trench Cutoff Walls

A soil–bentonite slurry trench cutoff wall, often known simply as a slurry wall, takes its name from the method of construction (Fig. 3.5). A trench is first excavated below the existing grade utilizing a liquid slurry of approximately 95% water and 4–6% bentonite by weight to maintain the trench stability. The bentonite–water slurry is a viscous fluid with a unit weight of approximately 64–70 pounds per cubic foot (pcf).

Trench collapse is prevented by the lateral pressure exerted by the liquid slurry on the filter cake balancing the active soil pressure. As a result of the higher fluid pressure within the trench, the slurry permeates the walls of the trench and, as bentonite is filtered from the slurry, a filter cake is formed on the sides of the trench. The filter cake is a very thin layer of fully hydrated bentonite. Detailed methods for calculating trench stability are available and are discussed elsewhere (Xanthakos, 1994; Nash, 1974).

The trench excavation is typically done using a backhoe with a modified boom to depths of approximately 55–60 feet and, below this depth, drag lines and clamshells are used (see Table 3.1).

The soil–bentonite slurry trench cutoff wall is completed by the placement of backfill which displaces the slurry (see Fig. 3.5). The soil–bentonite backfill is placed with a consistency of high slump concrete, which moves forward down the slope of previously placed backfill. The soil–bentonite backfill is

TABLE 3.1 Slurry Trench Excavation Techniques

Type	Trench Width (m)	Trench Depth (m)	Comments
Standard backhoe	0.5–2.0	10	Rapid, least costly
Backhoe with extended boom	0.5–2.0	16	Rapid and low cost (extended excavation boom on unmodified backhoe)
Modified backhoe with extended boom	0.5–2.0	20–30	Extended excavation boom; modified power and counterweights
Clamshell	0.3–2.0	75	Attached to crane
Dragline	1.0–3.0	50	Slow, for wide, deep excavations
Rotary	0.5–2.0	75	Slow, for boulders and rock
Specialty	Varies	Varies	For example, Hydrofraize (from Soletanche)

composed of well-graded soil, water, and bentonite. The bentonite content of the backfill may be as low as 1% or as high as 4–5%. The hydraulic conductivity of soil–bentonite backfill generally ranges from 1×10^{-7} to 1×10^{-8} cm/s.

Permeability of the completed cutoff wall can be affected by defects in the wall. Defects can result in a greater localized rate of contaminant migration as shown for soil–bentonite slurry trench cutoff walls in Fig. 3.6. Nonhomogeneous backfill results from improper mixing and includes lumps of unmixed soil and pockets of free slurry not fully blended with the soil. During backfill, slurry may become entrapped by free-dropping of backfill into the trench or as the backfill moves forward (Evans et al., 1985). The potential for slurry entrapment increases with increasing wall depth (McCandless and Bodocsi, 1987). If the trench is allowed to remain open for too long, sediment may accumulate and become trapped beneath the backfill. Cave-in of the trench sides or spalling from the formation into the excavated trench may also occur. There is also the potential for inadequate excavation of the key. Keying into rock is difficult and, unless measures are taken (such as preblasting along the alignment), cutoff wall effectiveness may be reduced due to flow beneath the wall (Trezek, 1986).

Potential long-term changes in the physical properties of the backfill must be examined. Such changes may result from (1) freezing and thawing; (2) wetting and drying; (3) desiccation of the backfill, and (4) chemical incompatibility. Cycles of freezing and thawing are known to increase the hydraulic conductivity of the clay mixtures (Othman and Benson, 1992; LaPlante and Zimmie, 1992; Zimmie and LaPlante, 1990; Zimmie et al., 1991; Zimmie, 1992). This can be prevented by a properly designed and constructed final

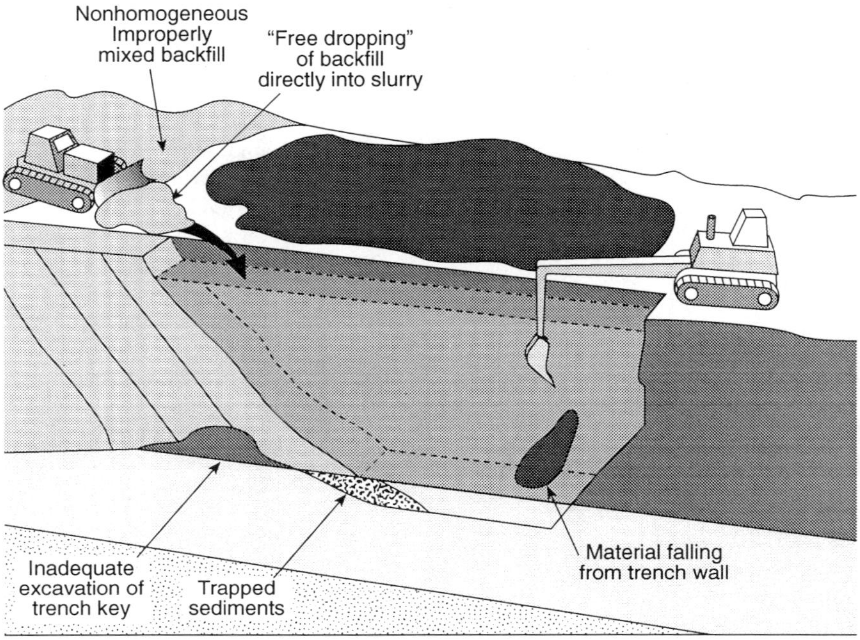

Figure 3.6 Potential defects in slurry trench cutoff walls.

cover over the cutoff wall. Cycles of wetting and drying are inevitable with a fluctuating water table. Drying can cause the bentonite to shrink resulting in an increase in hydraulic conductivity. Portions of the cutoff wall that are above the water table may desiccate, resulting in increases in hydraulic conductivity (see more detailed discussion in Section 3.4.3). Attention must also be given to chemical incompatibility which is covered in more detail in Section 3.5.2. Compatibility tests can be conducted to simulate a barrier wall in the presence of aqueous phase contaminants or nonaqueous phase liquids (NAPL). In the case of a contaminated environment, backfill employing well-graded materials, as described above, is recommended (along with site-specific compatibility testing).

3.3.2 Cement–Bentonite Slurry Trench Cutoff Walls

Cement–bentonite slurry trench cutoff walls are excavated under a head of slurry composed of water, cement, and bentonite and have been widely employed for groundwater control (Jefferis, 1981). The slurry is left to harden in the trench. As the cement hydrates, the combination of the bentonite, cement, water, and sand results in a hardened material with a consistency of stiff clay. Since excavated soils are not used as backfill, excavated materials require disposal. Cement–bentonite cutoff walls are generally limited to a hydraulic con-

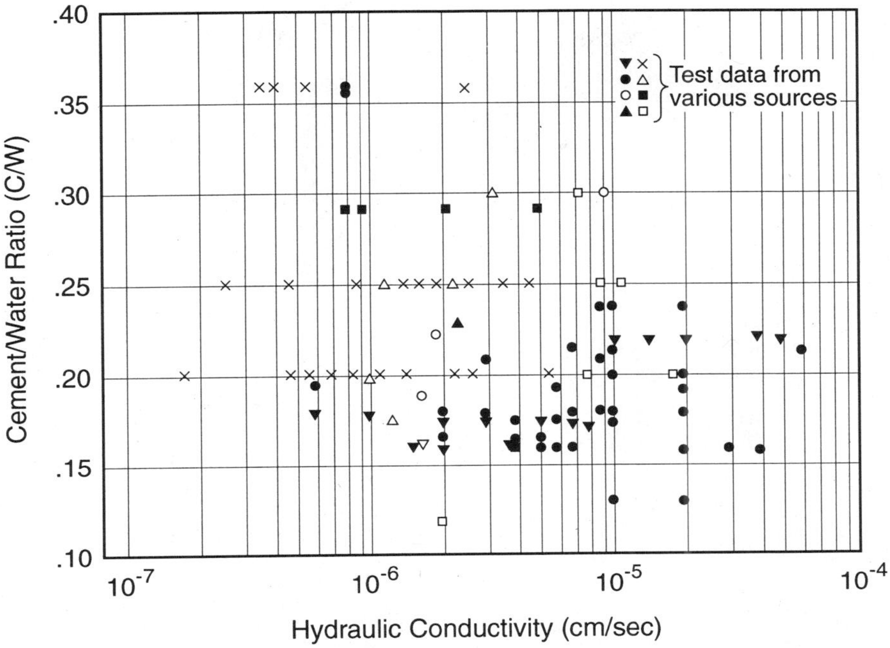

Figure 3.7 Cement–bentonite permeability test results (after Ryan, 1987).

ductivity of approximately 10^{-5}–10^{-6} cm/s, as shown in Fig. 3.7. The cement–water ratio is typically between 0.15 and 0.25, resulting in a cutoff wall that has a very low solids content; the wall is mostly water. Thus, cement-bentonite may be more prone to increases in hydraulic conductivity in response to contaminated fluids, although limited data in this regard are available. Arnould et al. (1991) developed grouts (Petrisol and Ecosol), that utilize standard bentonite and cement with the addition of siliceous materials, special clays, and chemical additives. Quantitative results of the effectiveness of these grouts as compared to traditional grouts have been presented. Generally speaking, cement–bentonite cutoff walls have found limited use for hazardous waste containment applications.

3.3.3 Plastic Concrete Cutoff Walls

Plastic concrete slurry trench cutoff walls may also be utilized for containment systems. Plastic concrete is a mixture of cement, bentonite, water and aggregate (unlike cement–bentonite, which does not include aggregate). Plastic concrete trenches are usually excavated in panels and the plastic concrete is placed in the slurry-filled trench using the tremie method to replace the bentonite-water slurry placed during excavation. For example, one project used a formulation of cement, pulverized fly ash, bentonite, and water to form the plastic

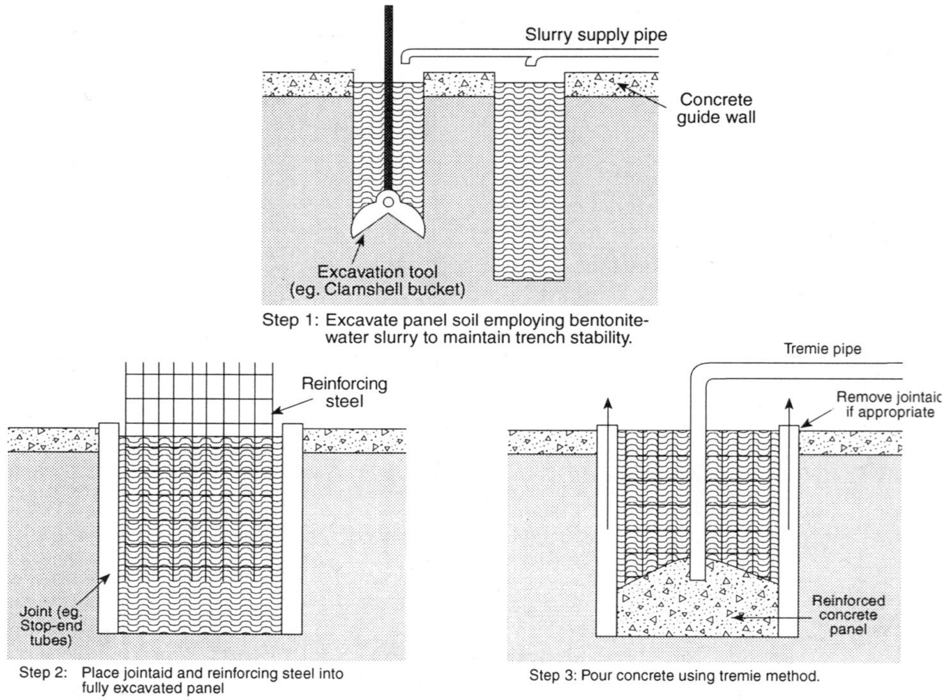

Figure 3.8 Diaphragm slurry trench cutoff wall.

concrete and employed the panel method of excavation to construct the cutoff to a depth of 160 ft below grade (Reina, 1987). Plastic concrete has relatively low hydraulic conductivity (10^{-6}–10^{-7} cm/s). The shear strength is significantly greater than soil–bentonite or cement–bentonite. Based upon limited data available to date, plastic concrete may be more contaminant resistant than other barrier wall formulations (Evans et al., 1987).

3.3.4 Diaphragm (Structural) Cutoff Walls

The deepest cutoff wall built in the United States (200 ft) was constructed as a diaphragm wall (ENR, 1990). In this technique, the trench is excavated under a head of bentonite-water slurry to maintain trench stability. However, the excavation is done in panels, each some 20 ft long, using a clamshell or similar excavating tool. Steel reinforcement is then set into the bentonite-water slurry. The slurry is then displaced by high slump concrete poured using the tremie method of concrete placement. A schematic of the method is shown in Fig. 3.8. Precast concrete panels have also been used to form the permanent diaphragm wall.

The diaphragm technique is widely used in applications requiring significant

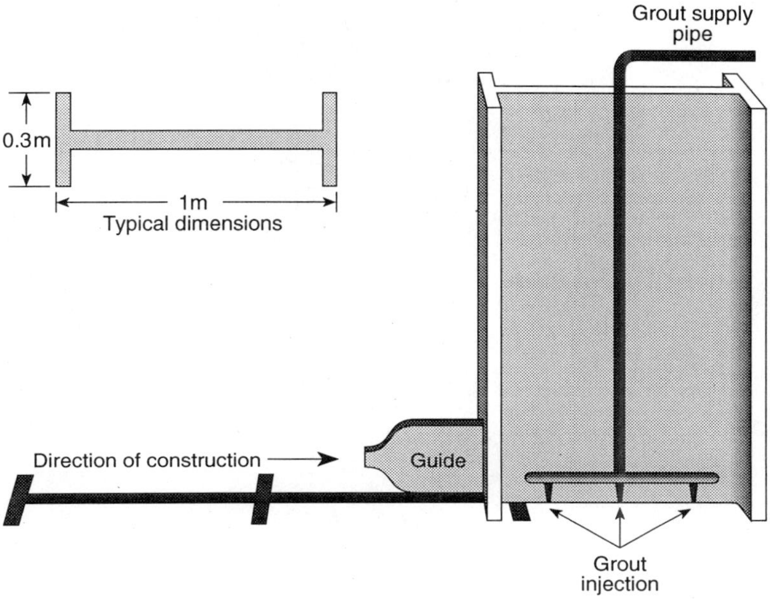

Figure 3.9 Vibrating beam.

wall strength, such as where the cutoff wall is to be part of a structural earth retention system (Boyes, 1975). The hydraulic conductivity is very low, typically 10^{-8} cm/s or lower. Due to the high cost, the diaphragm wall has not, as yet, been used in hazardous waste management applications.

3.3.5 Vibrating Beam Cutoff Walls

Vibrating beam cutoff walls have been employed for the control of contaminant lateral migration. The vibrating beam cutoff wall is constructed as shown in Fig. 3.9. The vibrating beam is a grouting method appropriate for soils. A vibratory pile driver is used to advance a specially modified H-beam into the subsurface. The pile is specially equipped with injection nozzles at the tip. During driving, grout is injected to lubricate the driven pile. As the beam is withdrawn, a void is left equal to the size of the beam. The void is filled with grout pumped in through injection nozzles as the beam is withdrawn. Subsequent beam penetrations are overlapped resulting in a continuous barrier with a typical thickness of 2–3 in. Cement–bentonite is the slurry most commonly employed. A special bituminous grout has occasionally been used. The principal advantage of this system is the elimination of the need to excavate potentially contaminated materials. Excavation of contaminated soils, as in the slurry trench methods described earlier, results in potential health and safety problems as well as added disposal costs.

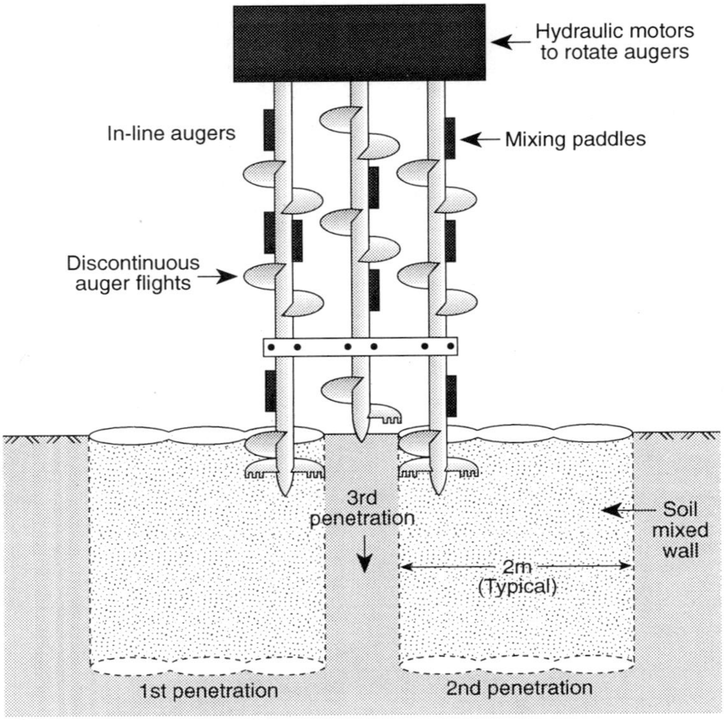

Figure 3.10 Deep soil mixing.

The integrity of the completed cutoff wall constructed using the vibrating beam method is subject to question. First, the cutoff wall is relatively thin, only a few inches. Second, the location of the beam tip is uncertain, particularly with deep penetration. Although vibrating beam walls have been constructed successfully to shallow depths (Leonards et al., 1985), it is difficult to penetrate dense soils to any significant depth. Furthermore, even sands that may be initially loose are densified by the vibratory action of the pile driver resulting in difficulty in subsequent penetrations. This may lead to the need for preaugering to achieve beam penetration (McLay, 1987).

3.3.6 Deep Soil Mixing

Deep soil mixing has been employed to construct vertical barriers. This technique was originally developed in Japan and has been recently employed in the United States. A special auger mixing shaft is rotated into the ground while simultaneously permitting the injection of bentonite and water, or cement, bentonite and water slurry. The installation sequence is shown in Fig. 3.10 and results in a column of treated soil. Multiple mixing shafts are usually employed. Reinforcing may be added to the treated soil columns if additional strength is required (ASCE, 1990). A continuous wall, typically from 20 to 36 in. wide, is obtained by overlapping penetrations. Bentonite is added to the mixed soil in the form of bentonite–water slurry. As a result, the bentonite content of the mixed soil is typically limited to about 1%. An innovative soil–cement mixed wall technique has also been described by Taki et al. (1990).

The advantages of the deep soil mixing technique are similar to those in the vibrating beam method. Health and safety risks are minimized by elimination of the need to excavate materials as required for conventional slurry trench cutoff wall techniques. The deep soil mixed wall can be constructed with added confidence since the wall is considerably wider than the vibrating beam wall. The hydraulic conductivity approaches 10^{-7} cm/s for deep soil mixed soil–cement walls (Yang et al., 1993).

3.3.7 Composite Cutoff Walls

Vertical barrier walls have been constructed using a combination of materials to form a composite barrier. The concept of including a geomembrane barrier within a slurry trench cutoff wall is one such composite. Two trade names are Gundwall® and Envirowall. Envirowall, as originally proposed, consisted of a trench excavated using the slurry method of excavation, followed by lining the trench with a high-density polyethylene (HDPE) geomembrane envelope. The geomembrane envelope is filled with sand to serve as a monitoring and/or withdrawal zone between the geomembrane layers (Ressi and Cavalli, 1985). Composite cutoff wall properties include lower hydraulic conductivity and increased contaminant resistance.

Although construction techniques vary between contractors, installation methods to date include:

1. The geomembrane sheet is mounted on an installation frame, lowered into the slurry-filled trench, disconnected from the frame, and the frame is withdrawn.
2. The geomembrane sheet is pulled into the slurry-filled trench using weights on the geomembrane bottom.
3. The geomembrane sheet is "driven" into the slurry-filled trench using a conventional vibratory pile driver.

It is reported that placement of a geomembrane in a cement–bentonite slurry trench cutoff wall decreases the overall hydraulic conductivity by two orders of magnitude. If high quality joining of geomembrane sheets can be achieved, this decrease in hydraulic conductivity can be four to five orders of magnitude (Jessberger, 1991).

In one case, the potential for the migration of methane gas from a landfill into adjacent land planned for residential construction was reduced by place-

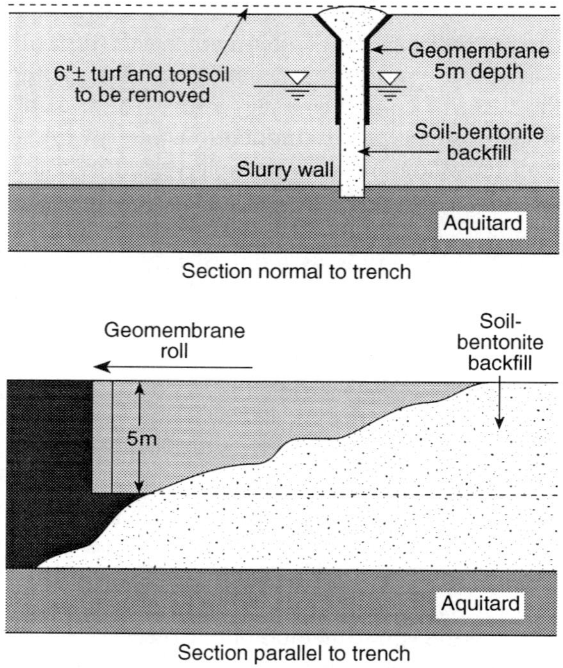

Figure 3.11 Composite cutoff wall.

ment of a slurry trench cutoff wall of soil–bentonite having a hydraulic conductivity of less than 10^{-6} cm/s. A 40 mil HDPE geomembrane liner was added to the upper portion of the wall to further reduce gas migration (GKN, 1989a). The geomembrane was considered necessary above the water table where the soil–bentonite is subject to wetting and drying cycles as well as potential desiccation drying and cracking. A schematic of the design is shown in Fig. 3.11.

3.3.8 Steel Sheet Pile Cutoff Walls

Steel sheet piling has found widespread application in conventional civil engineering construction where the retention of earth is coupled with the reduction of groundwater flow for excavations in the subsurface. Steel sheet piling has had limited application for environmental containment because steel sheet piling is subject to leakage through the grouted interlocks at rates higher than typically accepted for waste containment applications. Thus only limited discussion is provided here on steel sheet piling.

The principal advantage of steel sheet piling is its strength. Regardless of the hydraulic conditions, steel sheet piling will not hydrofracture; thus, under very high head conditions, the integrity of the cutoff can be maintained.

To reduce seepage through the interlocks, steel sheet piling has been used in conjunction with cement–bentonite on at least two occasions. For a nuclear power plant in Ohio, a cement–bentonite slurry trench cutoff wall was first excavated into the subsurface. Prior to the initial set of the cement–bentonite slurry, steel sheet piling was inserted into the cutoff wall. A composite wall constructed in this manner has high resistance to hydraulic fracturing under the projected head conditions. The impedance of leakage through the joint was achieved by grouting the joints with cement–bentonite. Although this solution was costly, the limited length of the required cutoff wall made a practical solution.

Until leakage through the interlocks of steel sheet piling can be controlled, it is likely that the use of steel sheet piling as vertical cutoffs in waste disposal applications will be limited. Also, the relatively high cost of steel sheet piling is a further disincentive to its use. However, sheet piling has been used in combination with soil–bentonite slurry where conventional soil–bentonite walls are not feasible, e.g., in rocky fill materials (McMahon et al., 1994).

3.3.9 Grouted Barriers

Grouts can be divided into two categories: particulate and chemical. Particulate grouts are composed of Portland Cement and/or clay and water, while the chemical grouts are composed of a chemical base, catalyst, and water or solvent. Examples of chemical grouts include sodium silicate, acrylate, and urethane. In order to get grout into the soil, the grout has to be mixed, a grouting tube has to be driven or drilled into the ground, and the grout has to be pumped into the ground.

Historically, grouts have been used for vertical barriers (cutoff walls or grout curtains) in dam construction by actually forming a wall in a soil formation or filling fractures in a rock formation. Grouted vertical barriers have also been used to contain buried wastes (Gazaway and Jasperse, 1992; Mitchell and van Court, 1992a). Customarily, such barriers have been keyed into impervious soil formations located below the waste for complete encapsulation.

a. Permeation Grouting. Permeation grouting involves filling the soil voids with grout. In order to fill the voids of a soil with grout, the permeability of the soil, viscosity of the grout, and size of particulates in the grout must be considered. The permeability of the soil is controlled by the grain size of the soil, which affects its average void size. The viscosity of the grout is dependent on the type of grout and how it is mixed. In general, particulate grouts have much higher viscosity than chemical grouts. In relation to soil permeability and grout viscosity, only soils with high permeabilities can accept particulate grouts, while soils with lower permeabilities can accept chemical grouts. Generally, soil hydraulic conductivities above 10^{-1} cm/s will accept particulate grouts and those above 10^{-3} cm/s will accept chemical grouts (Karol, 1990). It is very important to carefully choose a grout compatible for a given soil. Fig 3.12 summarizes the range of acceptable soil types for each type of grout.

The main concern when grouting to achieve low permeability is the complete filling of the soil void spaces with grout and the control of the lateral extent of grout penetration. Both of these tasks are very difficult and require skillful experience to control. Ueda et al., (1987) describe the application of an acoustic emissions technique during the grouting of rock masses and the use of such data in following the progress of the grouting.

There are two main methods of permeation grouting: point injection and sleeve pipe injection (or tube-a-manchette method). In the point injection method the casing is driven to full depth, then withdrawn to the desired depth and the grout is injected. In the tube-a-manchette method a sleeve pipe is placed in a grout hole and sealed in place by a clay–cement mixture. At 1-ft intervals the pipe has small holes through which the grout is pushed. The holes are covered by rubber sleeves, or manchettes, which act as one-way valves and open when the grout is pressurized (see Fig. 3.13). To inject grout, a double packer attached to a smaller diameter grouting pipe is inserted into a polyvinyl chloride (PVC) sleeve pipe centered on a sleeved hole, the pipe is pressurized and the grout is forced through the sleeved hole and into the soil.

The tube-a-manchette method is considered more suitable than point injection for the creation of a migration control barrier because the method allows the regrouting of a hole if the required watertightness was not achieved. The reuse option of the tube-a-manchette also permits the use of two different types of grouts: one (e.g., cement–bentonite) to fill the larger voids and another (e.g., acrylate) to fill the smaller voids.

Grout pumped at excessively high pressure can cause hydrofracturing of the soil. Hydrofracturing of the soil is preceded by an increase in grout pressure, followed by a sudden decrease in pressure as new grout channels are created by the fracture. Detrimentally, penetration of the grout into the fracture prevents the grout from filling the rest of the soil voids. To avoid hydrofracture, grout pressures should not exceed the soil fracture pressure. As a rule, pumping pressures should not exceed 1 pound per square inch (psi) per foot of soil overburden unless additional confinement than provided by the soil overburden exists. However, higher pressures have been successfully used since the above consideration does not account for the mobilization of shearing strength in the soil.

Permeation grouting to obtain a vertical barrier normally requires a minimum of two or three adjacent rows of grout columns to achieve the desired reduction in hydraulic conductivity. For three adjacent rows, normally the out-

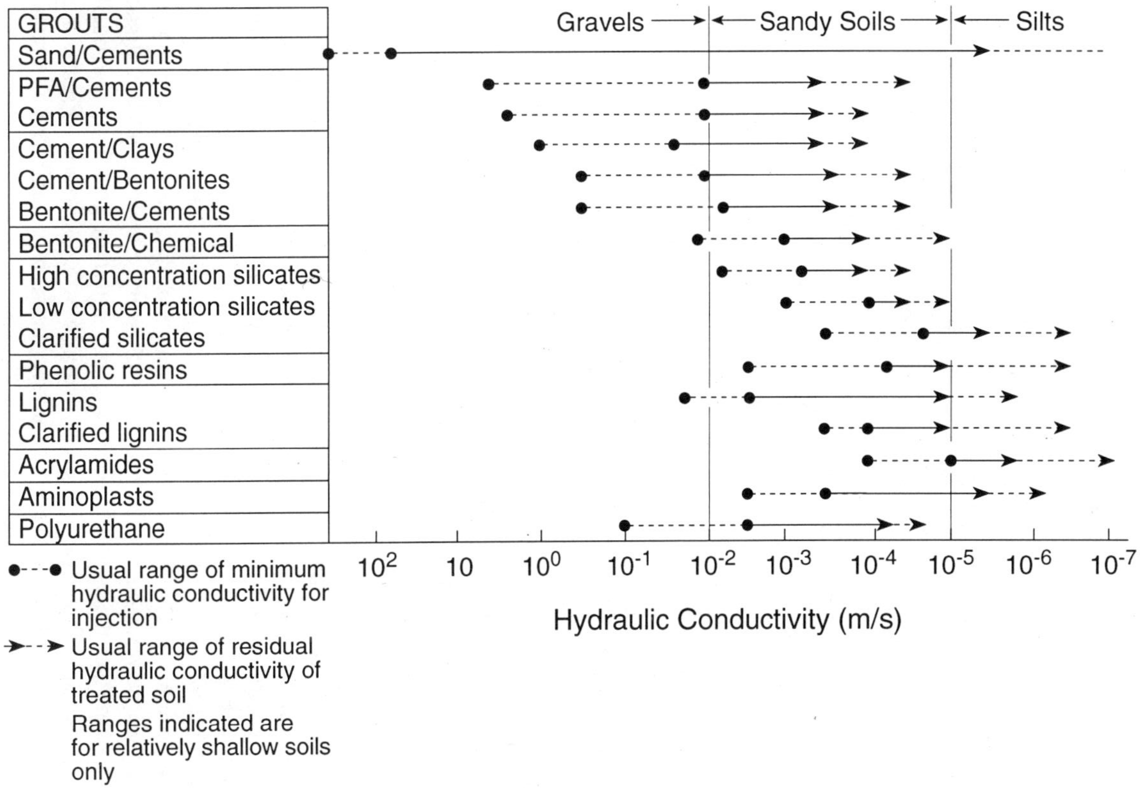

Figure 3.12a Relationship between grout type and soil permeabilities (after Karol, 1990).

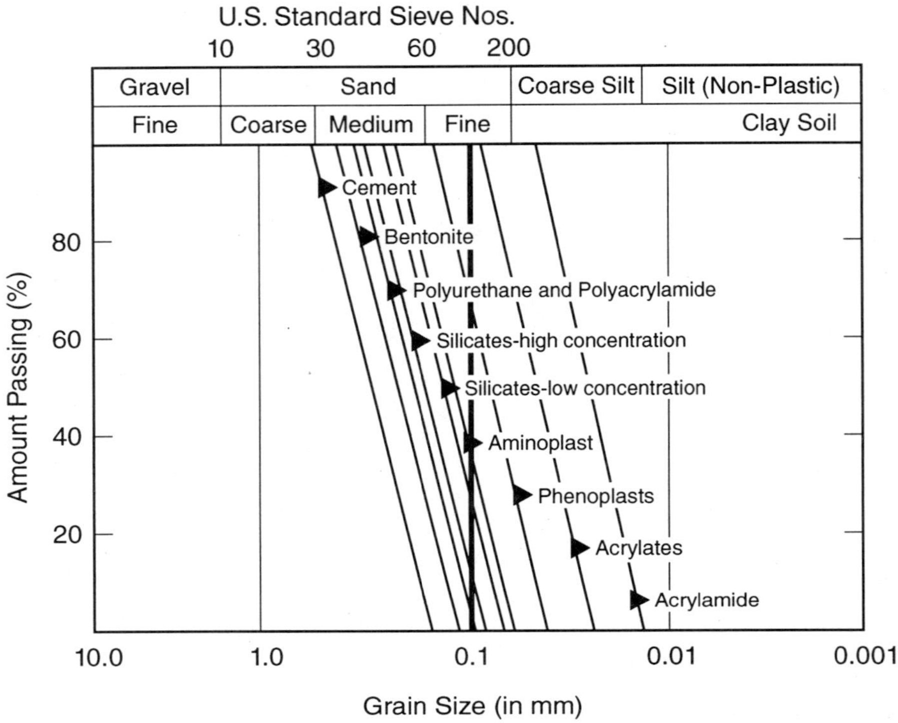

Figure 3.12b Relationship between grout type and grain size (after Karol, 1990).

side rows are grouted first so that all of the grout columns in one row are in contact, thus providing barriers on both sides of the center row. Within each of the outer two rows, every other grout column is grouted first, providing penetration barriers for the remaining grout columns in the row. Then the remaining grout columns are grouted in the outer two rows. Upon completion of the outer two rows, a center row of grout columns is placed between the outer two rows. Again, every other grout column is grouted first so that they touch the outer grouted rows. When grouted, the remaining columns fill the ungrouted space between the outer two rows of grout columns and the already grouted columns in the center row. Using this order of grouting, practically complete flow cutoff can be achieved within a soil formation. Figure 3.14 shows the geometric layout and order for a permeation-grouted low-permeability vertical wall.

When constructing a grout barrier, careful records of the pressure and volume of grout must be kept. Quality control is of the greatest importance when grouting a low-permeability barrier. If too small an amount of grout is used in one area of the barrier, it may be an indication that an ungrouted window in the barrier needs regrouting. If high pressure is not obtained while pumping

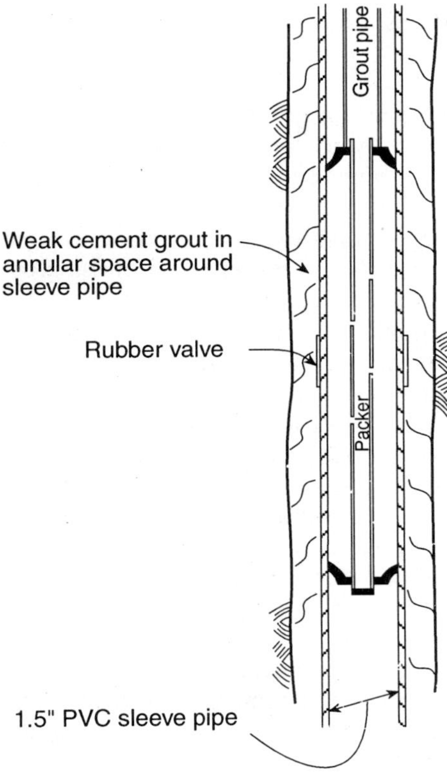

Figure 3.13 Tube-a-manchette (after Karol, 1990).

grout into the columns of the middle row, it may indicate that the outer two rows were not completely filled with grout. In order to regrout, it is convenient to use the tube-a-manchette method.

The hydraulic conductivity of a constructed grout barrier can be checked using field tests. Field tests can involve the use of piezometers on the upstream and downstream sides of the barrier and/or observation wells and dyes may

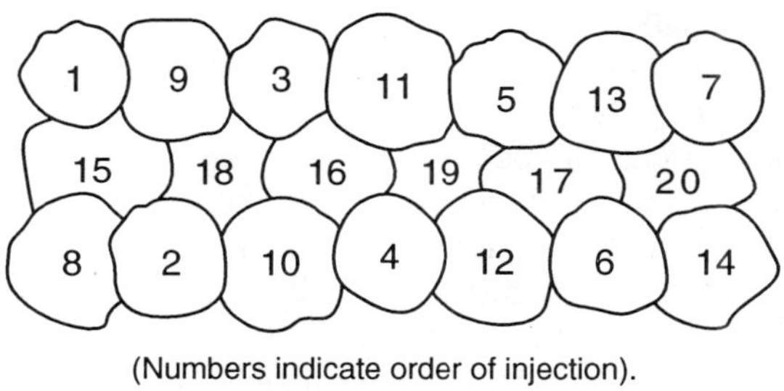

Figure 3.14 Geometric layout of injection grouted holes for low-permeability barrier.

also be used. Quality control considerations are discussed in greater detail in Section 3.8.

b. Jet Grouting. Jet grouting involves the use of grout alone, grout and air, or a combination of grout, air and water delivered by a small jet, or jets, in the drill rod at very high pressures, which often reach 5000–6000 psi. After advancing the drill rod to the desired depth, it is lifted and rotated as the jetted grout cuts away the soil and creates a large cylindrical hole. Portland Cement or cement–bentonite grouts are generally used when jet grouting. Cement grout mixes with the soil to form a soil and cement mixture (or soilcrete) column in the ground. The excess water and soil (spoil) is forced to the surface around the drill rod (Hayward-Baker, 1992; Mitchell and van Court, 1992a).

The methods of jet grouting commonly employed include: (1) the single-rod system, (2) the double-rod system, and (3) the triple-rod system. The number of rods refers to the number of passageways in the drill rod. Using the single-rod system, only a cement–bentonite grout is pumped out by the jet. Once the soilcrete is formed in the column, the excess soil is displaced to the surface. The single-rod system can make column diameters of 0.5–1.2 m in granular soil and 0.4–0.8 m in cohesive soils (Kauschinger et al., 1992). Of the three jet grouting methods, the single-rod system produces the smallest volume of spoil per column volume.

The double-rod system uses a combination of cement–bentonite grout and air pumped out by the jet. The addition of the compressed air enables the double-rod system to produce column diameters up to twice as large as the single-rod system, because the compressed air acts to hold back the groundwater and soil cut by the jet, preventing them from falling back into the jet stream. Also, the bubbling action of the compressed air helps to remove the soil from the cutting area and lift it to the surface (Kauschinger et al., 1992). The double-rod system forms a soilcrete mixture with a very high air content, which may increase the permeability of the grouted soilcrete column.

In the triple-rod system, which uses cement–bentonite grout, water, and air, soil removal is more complete and pure grout almost completely replaces the soil. Triple-rod systems can produce column diameters in granular soils between 0.5 and 3 m and column diameters in cohesive soils between 0.5 and 1.5 m (Kauschinger et al., 1992). Large diameters can be achieved since the triple-rod system actually employs two jets: a top jet for the air and water, which performs the cutting and spoil removal, and a bottom jet that fills the column with the grout.

Nontoxic cement or cement–bentonite grouts can be used when jet grouting, offering the added advantage of resistance to chemical leachates and relatively low cost. Cement grouts with microfine cement and other admixtures (e.g., bentonite or microsilica) have initial hydraulic conductivities in the range of about 10^{-7}–10^{-10} cm/s (Weaver et al., 1990; Bowers et al., 1988). Research is being conducted to find better mixture or materials to improve the jet grouting procedure (Burke, 1993; Welsh, 1993). One disadvantage to jet grouting is that the only exit for the displaced soil cuttings and water is the drill hole opening at the ground surface. If this opening is blocked, excess pressures can build up, and hydrofracture may occur. As stated earlier, pumping pressures must be monitored carefully. A second disadvantage is the potentially large volume of spoil produced. The spoil from a waste site usually has to be cleaned and redeposited on site and can greatly increase the cost of jet grouting. However, the quantity of spoil can be controlled (Burke, 1993; Welsh, 1993). To create a vertical wall around the waste, a single row of jet grouted columns can be used with carefully controlled spacing so that adequate overlap of the columns is ensured. The vertical wall (grout curtain) needs to be keyed into a natural clay or rock. If a grouted bottom barrier is to be constructed, the side curtain is constructed first with the bottom seal keyed into the side curtain. Jet grouted barrier walls can be constructed to depths of approximately 36 m (Burke, 1993; Welsh, 1993). Beyond this depth, vertical alignment of the drilled columns is difficult to control and the integrity of the barrier grout curtain cannot be ensured.

c. Soilsaw™ Barrier Wall. The Soilsaw™ vertical barrier placement system is a new technology being developed by Brown & Root Environmental in Houston, TX, in cooperation with the Army Corps of Engineers (Corps) and the Department of Energy (DOE) (Dougherty, 1993).

The Soilsaw™ is an *in situ* method of constructing soil–bentonite or cement–bentonite slurry walls with low hydraulic conductivity. The process converts undisturbed soil into a viscous, mortar-like material of specified depth and width in a single step. Wall properties typically exceed those found in slurry walls constructed using conventional construction techniques.

The system consists of a heavy pipe, called a beam, mounted on a specially modified, tracked excavator. The beam is equipped with a high-pressure fluid conduit that terminates in multiple jets placed at regular intervals down the length of the beam. The system uses Halliburton high-pressure pumps to accelerate a specially blended cement–bentonite slurry to a very high kinetic energy through these jets. The high-velocity slurry impacts the soil, disrupting the soil structure and creating an intimate blend of soil and slurry to form the *in situ* slurry wall.

Once the slurry contacts the soil, it is incapable of conveying any pressure to the walls of the slurry–soil trench other than that due to the hydrostatic head of the column of fluid in the ground. This feature makes the system ideal for reinforcing structures such as earthen levees, which cannot tolerate excessive internal pressures.

This construction methodology produces no excavated material during slurry wall construction and during wall placement in contaminated soils no waste is generated due to excavation. Because the process is continuous, on-site construction time is minimized and disruption to normal operations is limited.

Operations are reasonably clean, since bentonite or cement–bentonite grouts

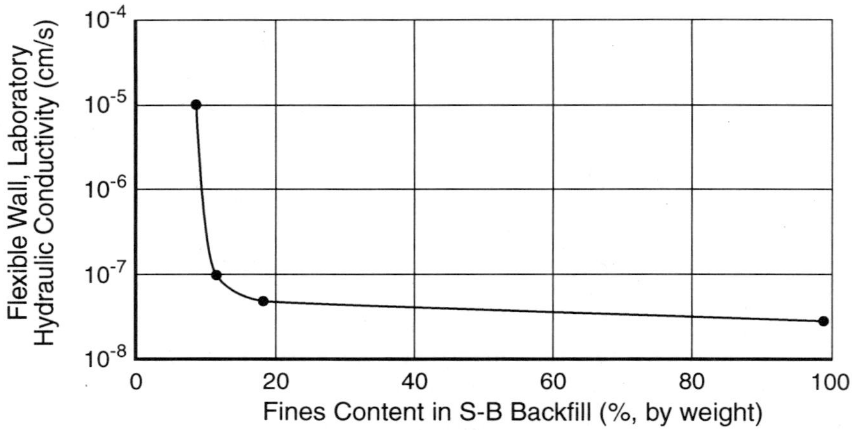

Figure 3.15 Relationship between fines content and hydraulic conductivity (Evans, 1994).

are mixed in a special batch plant on site and transferred via a low-pressure pipeline until the material is accelerated to the jets during wall construction.

3.4 SLURRY WALL PERFORMANCE FACTORS

3.4.1 Effect of Grain Size Distribution

Early data revealed that the type and nature of the fines fraction of the soil influence the hydraulic conductivity of soil–bentonite backfill (D'Appolonia, 1980). An increase in the fraction of the soil finer than No. 200 sieve generally results in a decrease in the hydraulic conductivity. Data taken from recent design studies for a soil–bentonite (SB) cutoff wall are shown in Fig. 3.15, demonstrating that hydraulic conductivity decreases with increasing fines content. The data demonstrate the use of natural fines to achieve low-hydraulic conductivity without enriching the mix with dry bentonite. For this particular study, 20% fines (material passing the No. 200 sieve) from a clayey borrow source added to the base soil of about 20% gravel, 70% sand, and 10% silt, resulted in a hydraulic conductivity of 5×10^{-8} cm/s, similar to the backfill using virtually 100% plastic fines from the clayey borrow source (3×10^{-8} cm/s).

Although there is clearly a relationship between fines content and hydraulic conductivity for these particular materials, the relationship may not be generalized for design purposes since the data shown in Fig. 3.15 are for specific soils from a specific site. In fact, data presented by Ryan (1987) in updating an earlier published relationship, demonstrated that it is not possible to assure a specified hydraulic conductivity by simply choosing a fines content from a

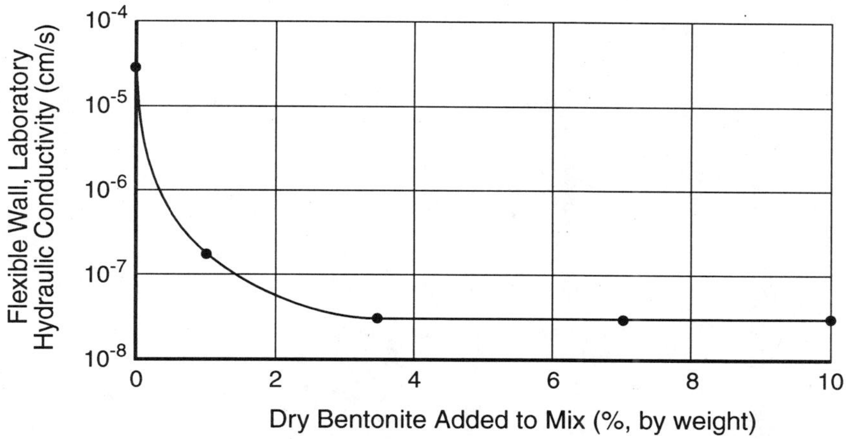

Figure 3.16 Effect of bentonite content on Hydraulic Conductivity of a soil–bentonite backfill (Barvenik, 1992).

published relationship. A specific relationship needs to be developed for each site.

3.4.2 Effect of Bentonite Content

Figure 3.16 shows the relationship between hydraulic conductivity and bentonite content for a particular project (Barvenik, 1992). For this particular mix, the minimum hydraulic conductivity was found at a bentonite content of about 3%. Relationships such as this may be developed for each project, but data from many projects (see Fig. 3.17) demonstrate that there is no general correlation of hydraulic conductivity to bentonite content (Ryan, 1987). The same conclusion can be drawn for cement–bentonite cutoff walls.

3.4.3 Influence of Fluctuating Water Table

A principal purpose of a cutoff wall is to impede the flow of groundwater (and the associated transport of contaminants in environmental applications). Considerable effort is made to hydrate the bentonite in the cutoff wall in an effort to achieve the desirable hydraulic conductivity. Furthermore, it is common to have a portion of the wall remain permanently below the water table, another portion permanently above the water table, and a portion that may be in contact with a fluctuating water table. Unfortunately, only limited information on the long-term performance of cutoff walls is available.

In one recent study, measurements of permeability were made for a soil–bentonite cutoff wall having one portion constructed in 1981 and another constructed in 1987. The two portions were joined to form a vertical cutoff surrounding a wet ash handling facility, which maintained essentially constant

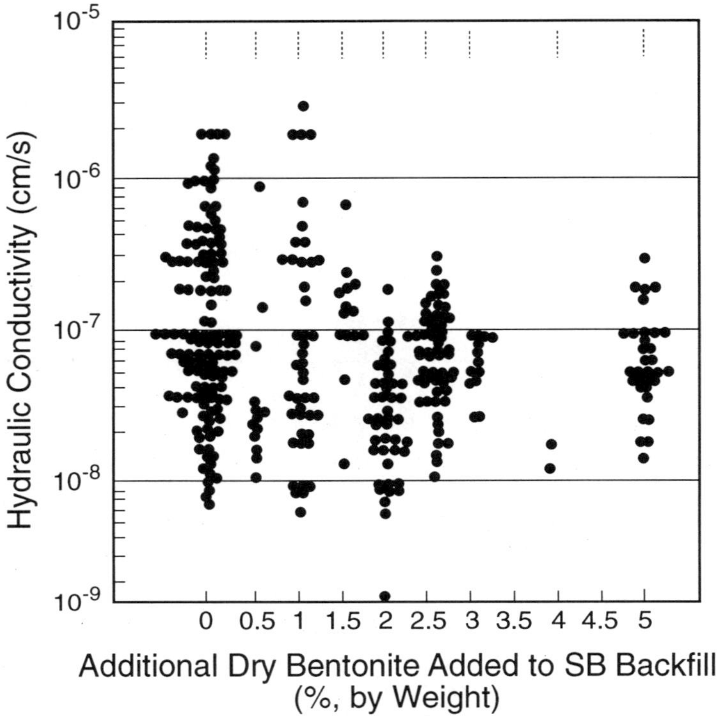

Figure 3.17 Soil–bentonite backfill laboratory test results (Ryan, 1987).

water levels throughout the year. Shelby tube samples were obtained above and below the water table for each portion of the cutoff wall and hydraulic conductivity tests on these samples were conducted. The results are shown in Table 3.2 (Cooley, 1991). The data show higher hydraulic conductivities for those samples above the water table than for those below the water table. The potential for "rehydration" was investigated by repeating the hydraulic conductivity tests after 16 days of backpressure saturation. No noticeable changes in the hydraulic conductivity were observed from the values presented in Table

TABLE 3.2 Effect of Water Table and Wall Age on Hydraulic Conductivity of Soil–Bentonite

Construction Date	Position with Respect to Water Table	Hydraulic Conductivity (cm/s)
1981	Above	1×10^{-3}
1981	Below	1×10^{-8}
1987	Above	6×10^{-6}
1987	Below	1×10^{-7}

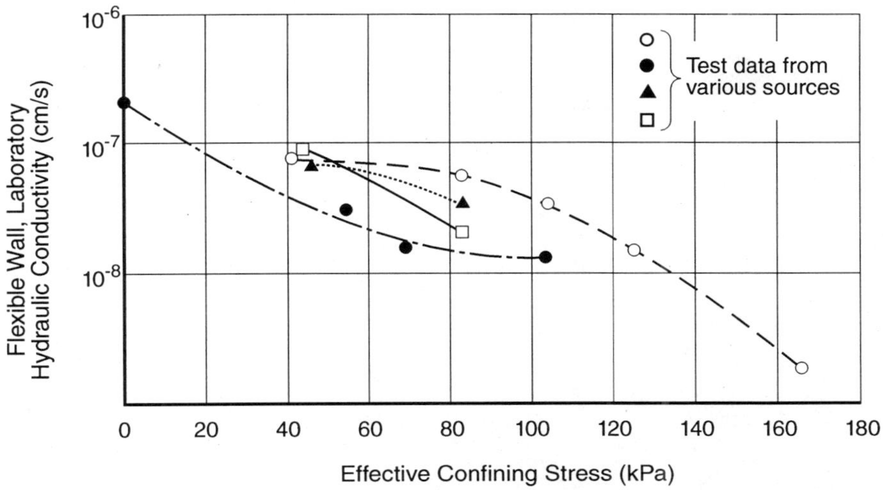

Figure 3.18 Effect of consolidation pressure on hydraulic conductivity of soil–bentonite backfill (data from McCandless and Bodocsi, 1988; Day, 1992).

3.2. The findings from this study suggest that soil–bentonite materials may need to be continuously saturated, otherwise irreversible increases in hydraulic conductivity may occur.

3.4.4 Effect of Consolidating Pressure

The hydraulic conductivity of vertical barrier materials decreases as the effective consolidating pressure increases. This relationship can be expected based on theoretical considerations of decreasing void ratio with increasing effective stress. Using data from several studies on soil–bentonite backfill, Fig. 3.18 presents relationships between effective consolidating pressure and hydraulic conductivity (McCandless and Bodocsi, 1988; Day, 1992).

3.4.5 Field Stress Conditions

Since the hydraulic conductivity is stress dependent, the state of stress in the field needs to be known for proper interpretation of hydraulic conductivity obtained from laboratory studies. Laboratory models and field data obtained to date on soil–bentonite slurry trench cutoff walls indicate that stress increases with depth at a rate less than the usual hydrostatic increase rate (McCandless and Bodocsi, 1987; Evans et al., 1985d). Analytical approaches reveal similar findings (Sweidan, 1990). Data on a fully instrumented soil–bentonite cutoff wall measuring total and effective stress with depth have not been found in the published literature. The soil–bentonite backfill is quite compressible compared with the relatively rigid trench sidewalls and, as a result, the consoli-

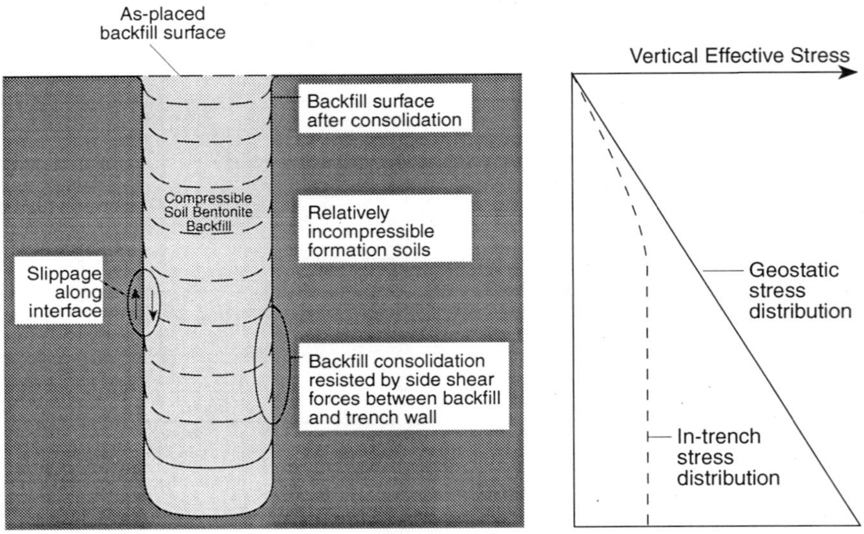

Figure 3.19 Slurry wall state-of-stress.

dation of the backfill is restricted by the friction at the trench–backfill interface. This action has been termed arching by some authors (Millet et al., 1992). Figure 3.19 schematically portrays the phenomenon.

An investigation of an existing soil–bentonite slurry trench cutoff wall was undertaken to further study this arching phenomenon (Cooley, 1991). The cutoff wall was built in two portions at two different times. At the time of testing one portion was 4 years old and the other was 10 years old. Shown in Fig. 3.20 is a plot of water content versus depth. If the soil–bentonite consolidated hydrostatically, the water content would be expected to decrease with depth (in response to increasing effective stress and corresponding decrease in void ratio) in the zone below the water table. The data show otherwise; generally indicating a constant water content with depth. It can also be observed that there is little difference in the water content profiles between the 4- and 10-year-old cutoff walls.

Shelby tube samples were obtained during the course of these investigations and subjected to consolidation testing; the results are summarized in Table 3.3. Knowing the stress history of these materials, it is reasonable to assume that they are normally consolidated or nearly so. They have not been subjected to significant surcharge loadings or fluctuating water levels, which could cause overconsolidation. They have been in place long enough to fully consolidate under the applied stress. Assuming a hydrostatic stress distribution, the over consolidation ratio (OCR) for each sample is shown in Table 3.3. The computed values range from 0.17 to 0.63. These values imply that the materials have consolidated to equilibrium under stress less than expected from a hydrostatic stress distribution. It is evident that the effective stress consolidating the

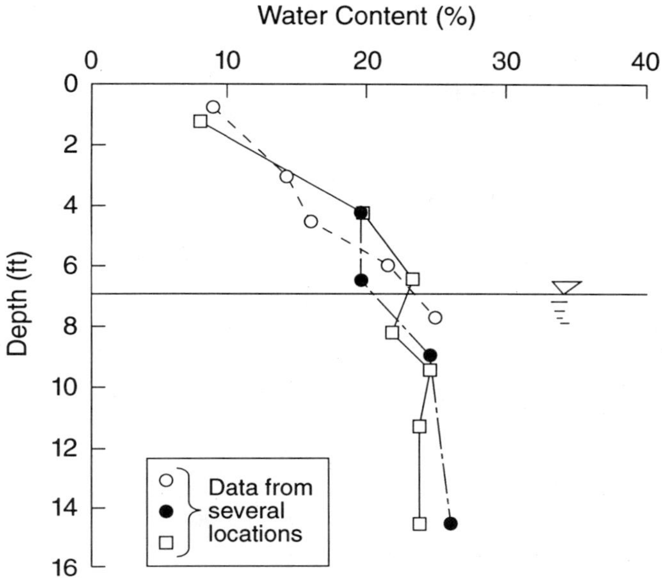

Figure 3.20 Water content versus depth.

materials is substantially less than that expected from a hydrostatic distribution, thus providing further evidence for the "less-than-hydrostatic" stress distribution in the cutoff walls.

Field studies using the flat plate dilatometer and analytical studies employing finite element methods (Duncan et al., 1984) were undertaken (Evans, 1993). The dilatometer and its associated data reduction program result in several "outputs" based on theoretical calculations and engineering correlations. For a wall tested on the same day as backfill placement, the engineering classification was "mud". For both the 10- and 4-year-old walls the classification was also generally given as "mud". The dilatometer was not able to discern much difference between material varying from just placed to in-place for 10 years. Using the correlation for K_0 from Brooker and Ireland (1965), coupled with triaxial shear strength tests, estimates of vertical stress were made from the dilatometer data. (K_0 is the ratio between the horizontal and vertical effective stress for a soil at rest that is not loaded.) Dilatometer profiles were also obtained at different orientations relative to the axis of the cutoff wall. These data are similar indicating little directional difference, at least within the capability of the dilatometer to distinguish anisotropy in these soft materials (Lutenegger, 1988). The results from finite element analysis, dilatometer data, consolidation data (assuming the measured preconsolidation pressure is equal to the vertical effective stress) and theoretical calculations of effective stress are shown in Fig. 3.21. The effective stress distribution is less than hydrostatic.

TABLE 3.3 Consolidation Test Results

Sample	Depth (ft)	Preconsolidation Pressure (psi)	Theoretical Preconsolidation Pressure	OCR	Compression Index	Modified Compression Index	Recompression Index	Modified Recompression Index
BC-2c ST No. 1	7.0	2.50	6.32	0.40	0.141	0.088	0.010	0.006
BC-2c ST No. 2	9.0	4.31	8.13	0.53	0.178	0.108	0.014	0.009
BC-2c ST No. 3	14.5	8.19	13.09	0.63	0.241	0.147	0.009	0.006
BC-3b ST No. 2	9.5	1.94	8.31	0.23	0.186	0.110	0.026	0.015
BC-3b ST No. 3	13.0	1.94	11.38	0.17	0.163	0.097	0.012	0.007

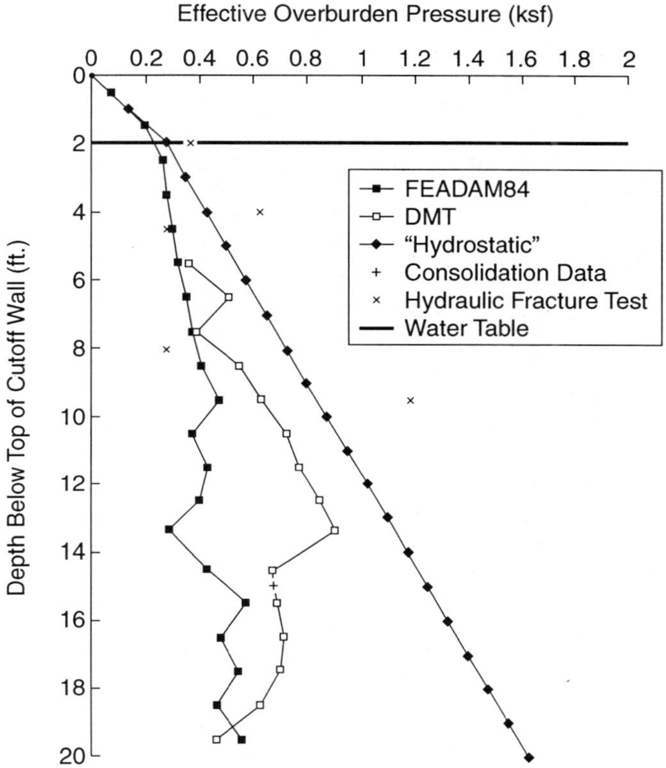

Figure 3.21 Vertical stress at the center of a soil–bentonite slurry wall.

It is recognized that the field situation was imperfectly modeled by the finite element efforts undertaken for this study. The program cannot simulate the construction process of excavation through a slurry and subsequent displacement of the slurry with soil–bentonite backfill. No properties for the embankment soil were available, so typical parameters were chosen from the literature for this soil. Soil parameters for the soil–bentonite backfill were based on consolidation and triaxial shear strength data and engineering correlations in the literature. It is also noted that the time rate of consolidation cannot be accounted for in this program. Finally, and perhaps most importantly, at the junction of the two soils (soil–bentonite and trench walls), a common nodal point is shared. Since elements cannot slide relative to nodal points (only deform or expand), this constraint probably affected the predicted stresses. Despite the limitations, it is useful to examine the results of these efforts.

Based upon the information available to date for soil–bentonite walls, the effective stress distribution with depth depends on: (1) wall thickness, (2) compressibility of backfill material, (3) backfill–trench wall interface friction; (4) backfill density, and (5) Poisson's ratio. Arching, and the associated reduction in effective strengths at depth, can be minimized by increasing the wall thickness and reducing the difference in compressibility between the backfill material and the native soil.

3.4.6 Hydraulic Fracturing Effects on Hydraulic Conductivity

The nature and potential for hydraulic fracturing in soil–bentonite slurry trench cutoff walls can be misunderstood. Handbook guidance quotes a rule of thumb of 1 psi per foot of wall depth as "safe against hydraulic fracturing" (Spooner et al., 1984). That is, if pore water pressure in excess of hydrostatic at various depths below the ground surface is less than 1 psi for each foot of depth, hydraulic fracturing is unlikely to occur. Alternatively, a width of 15–23 cm (0.5–0.75 ft) per 3 m (10 ft) of hydrostatic head is cited (Case International, 1982).

In waste containment applications, this guidance has been used to calculate the maximum drawdown from within the barrier. The guidelines were originally developed for cases such as excess pore water pressure in a slurry wall beneath the core of a dam. In such an application, the excess pore water pressure could exceed the minor principal total stress within the cutoff wall, reducing the effective stress to zero and resulting in hydraulic fracturing. However, dewatering from within a cutoff wall cannot induce excess pore water pressures above those existing prior to dewatering; therefore, hydraulic fracturing will not result from dewatering.

Hydraulic fracture testing can be used to study the state-of-stress in cutoff walls. Hydraulic fracture tests at a depth of 20 ft on the 4-year-old cutoff wall (described previously) resulted in hydraulic fracturing at an excess head of about 4 ft. This corresponds to an excess fluid pressure of about 0.25 kips per sq. ft. (ksf) and to an effective stress of zero in the soil. Thus the total stress on the plane of fracturing is about 0.25 ksf. Assuming K_0 conditions (K_0 = 0.37 from the Brooker and Ireland correlation based on data from site specific triaxial testing), the vertical stress would be estimated to be 0.68 ksf. This datum point is also plotted in Fig. 3.21 and is consistent with the other data, adding further evidence that the state-of-stress in soil–bentonite cutoff walls is less than hydrostatic.

Polozov and Popov (1985) studied the problem of interconnecting fractures employing hydrofracturing followed by grout injection. Mori and Tamura (1987a) investigated hydrofracturing phenomena in soils. The hydrofracturing pressure of sandy soils is strongly related to the permeability as well as the confining pressure, and this pressure in a cohesive soil is the sum of the minimum confining pressure and the unconfined compression strength of the soil. Mori and Tamura (1987b) investigated hydrofracturing pressures in vertical boreholes for six kinds of cohesive soils. The shear failure, rather than the tensile failure, near the borehole initiates the hydrofracture in cohesive soils. Dvorkin et al. (1992) reported results from their investigation of the dynamics of a viscous compressible fluid in a fracture. Rutovist et al., (1992) reported results of a theoretical and field investigation of the coupled hydromechanical behavior of fractured rocks. Their results validate the soundness of the conceptualized constitutive relations governing the coupled hydromechanical behavior of rock masses.

3.5 TEST METHODS AND COMPATIBILITY

3.5.1 Test Methods

It is not within the scope of this book to provide a comprehensive summary of literature relating to test methods for hydraulic conductivity. The reader is referred to other sources for a more complete treatment of the subject (Bowders et al., 1986; Daniel et al., 1984; Dunn and Mitchell, 1984; Lambe and Whitman, 1979; Olson and Daniel, 1981).

Studies have shown that the type of permeameter has little effect on the hydraulic conductivity of a compacted clay when measured in a laboratory using water as the permeant (Daniel et al., 1985); especially when the impact of effective stress is considered (Manuel et al., 1987). Sidewall leakage, and accompanying sample shrinkage, contributes to the magnitude of the hydraulic conductivity increases in studies using fixed wall permeameters with concentrated organic permeants. Regardless of the type of permeameter used, the ability to maintain constant influent chemistry throughout the test is necessary (Evans and Manuel, 1985). There are a number of testing parameters that can affect the outcome of a laboratory compatibility test (Evans and Fang, 1983, 1988; Zimmie et al., 1981).

Nordquist et al. (1986) describe the results of measuring the hydraulic conductivity of clay liners both in the field and in the laboratory. Averages of field and laboratory results were within one order of magnitude and indicate the need for careful field quality control methods. Grube (1990) reviewed the different problems associated with measuring the performance of clay containment barriers. An evaluation of the hydraulic performance of barrier dikes and cutoff walls is more complex than in analyzing the behavior of normal undisturbed soil systems. Anderson et al. (1991) described a calibration chamber suitable for the preparation of uniform clay beds in which the performance of full-size field test devices may be studied. McBride and Baumgartner (1992) described an inexpensive slurry consolidometer that uses porous polyethylene as the permeable barrier for sample dewatering and permits monitoring of sample pore-water pressures less than 100 kPa with a portable pressure transducer. Tan et al. (1992) described a practical method of measuring the *in situ* slurry density profile with depth using a submersible gamma source, backscatter-type nuclear density gauge.

It is also difficult to clearly establish the compatibility test duration. Although passing of two or three pore volumes has been the generally accepted practice, it is not clear that this criterion is always suitable. Studies of clays

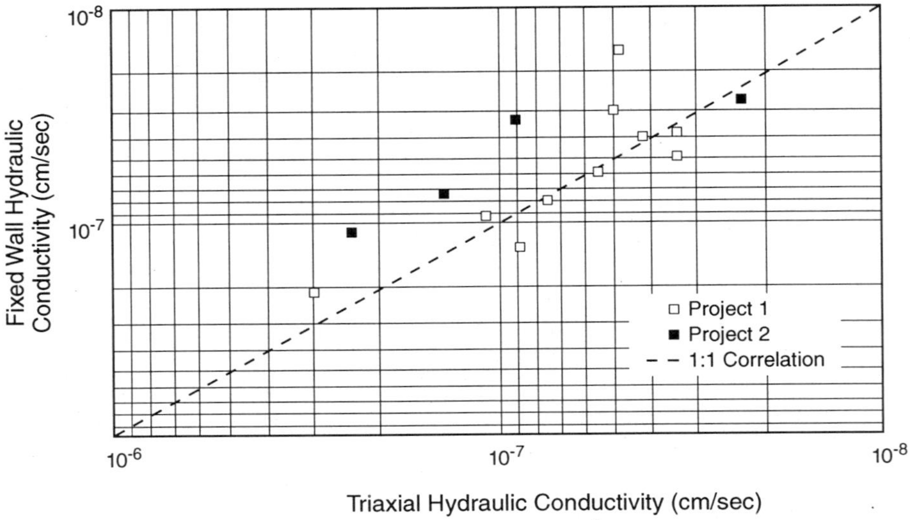

Figure 3.22 Permeameter test results (data from Day, 1992; Barvenik, 1992).

permeated with concentrated organics have generally shown dramatic hydraulic conductivity increases within two pore volumes. However, where dissolutioning of the soil structure is occurring, the time required to establish a new equilibrium hydraulic conductivity may be much greater than the time required to pass two pore volumes. Some data have shown that the number of pore volumes may be less important than the time of exposure (Bodocsi et al., 1987). Pierce and Witter (1986) suggest one pore volume of fluid with the condition that the slope of hydraulic conductivity versus pore volume displacement be essentially zero. Alternatively, Bowders (1988) suggests two pore volumes of flow and that the influent chemical concentrations be essentially the same as the effluent chemical concentrations.

It has been shown that hydraulic conductivity testing can be accomplished relatively rapidly on-site using a rigid wall permeameter to achieve results comparable to triaxial tests (GKN, 1989b). Typical data are shown in Fig. 3.22. The determination of hydraulic conductivity must be at the same gradient and consolidation pressure as expected in the field since soil–bentonite backfill hydraulic conductivity is stress dependent (McCandless and Bodocsi, 1988). In general, rapid field permeability tests are conducted in an American Petroleum Institute (API) filter press rigid wall permeameter.

3.5.2 Compatibility

A major concern in the application of vertical cutoff walls to site remediation is the compatibility of the barrier material with the site specific contaminants. This section presents results reported in the literature and provides assessment of the findings with respect to the compatibility of vertical barrier wall materials. Readily available information regarding clay mineralogy and clay–water–electrolyte systems is not presented here. The reader is referred to existing publications (Grim, 1968; Mitchell, 1993; Van Olphen, 1977) for detailed information.

Widespread concerns regarding the potential incompatibility of clayey materials with subsurface contaminants began to surface in the late 1970s and early 1980s. In a paper describing slurry cutoff walls for control of hazardous wastes, Ryan (1985) does not present any compatibility data and states that "We have yet to find a leachate whose effect on the soil–bentonite backfill cannot be counteracted by relatively minor changes in the constituents." In a paper describing the use of slurry walls, D'Appolonia (1980) presented data indicating that certain inorganic chemicals could cause increases in the hydraulic conductivity of the filter cake. Early work, published while studies were still in progress, indicated the potential for large increases in hydraulic conductivity as a result of permeation of clayey soils with concentrated organic liquids. As a result of this concern, a number of investigations were undertaken to further investigate the phenomenon. However, most of the studies dealt with compacted clays, which are considerably denser and have much lower water contents than slurry walls.

Brown and Anderson (1983) found that permeation of compacted clays with concentrated organic fluids can lead to increases in hydraulic conductivity from one to three orders of magnitude. They assessed their findings in light of the accepted colloidal models of soil behavior and concluded that the behavior was consistent with these models (such as Gouy–Chapman). The studies were performed using rigid wall, compaction mold permeameters. Anderson et al. (1985) found similar dramatic increases in hydraulic conductivity when testing bentonite slurries with concentrated organics (xylene and methanol) in double ring fixed wall permeameters.

Acar et al. (1985) also found increases in the hydraulic conductivity of compacted kaolinite when exposed to concentrated organics. Importantly, these studies examined the influence of confining pressure and fluid concentration on changes in hydraulic conductivity. For permeation with concentrated organic fluids, dramatic increases in hydraulic conductivity were observed for tests done in rigid wall permeameters but, as the confining pressure increased in tests using flexible wall permeameters (triaxial cells), the magnitude of the hydraulic conductivity increases decreased. Stated another way, the ratio between the final and initial hydraulic conductivity decreased as the confining pressure increased. The study also found that all tests conducted using organics at low concentrations resulted in slight decreases in hydraulic conductivity. The hydraulic conductivity data, as well as indicator data (Atterberg limits and free swell) were found to be consistent with changes expected from variations in surface interaction forces between the colloidal clay particles.

In another study, it was found that the hydraulic conductivity ratio (the ratio between the hydraulic conductivity with the contaminant to that with water) increased with decreasing activity, whereas, volume decreases were greater with increasing activity (Acar and D'Hollosy, 1987). Activity is defined as the plasticity index normalized with respect to the clay fraction; high activity clays, such as bentonite, are thus subject to greater ranges of shrinking and swelling than low activity clays, such as kaolinite. The results indicate that while soil fabric changes are important in controlling hydraulic conductivity, at increasing activity the volume change becomes the controlling factor.

Studies of soil–bentonite materials in triaxial permeameters demonstrated significant increases in hydraulic conductivity with concentrated organic fluids, whereas, virtually no effect was observed for the same fluids dissolved in water at low (up to 30,000 ppm) concentrations (Evans et al., 1985b and c).

The effect of inorganic permeants on bentonite was examined by Alther et al. (1985) employing filter press and cracking pattern tests. These investigators found that there was an increase in hydraulic conductivity with increasing electrolyte concentration. They also found that divalent cations had a greater impact than monovalent cations. These data were generally consistent with that expected from an examination of the colloidal behavior as described by the Gouy–Chapman model. When all else is held constant, the permeability should also increase with smaller hydrated ions. The data were inconclusive in this regard.

In another study of inorganic permeants, the pH of tap water was varied using hydrochloric acid and sodium hydroxide to achieve pH ranges from 1 to 13 (Lentz et al., 1985). Three clays were tested using these permeants at gradients of 400–500, all without any detrimental effect. Since acids would be expected to result in permeability increases it is necessary to question whether these data are an artifact of the test conditions. Jefferis (1992) warns that tests must be carried out for sufficient time to allow for the full effects of reaction, particularly where relatively few pore volumes are permeated under high gradients. An examination of Fig. 3.23 demonstrates that, if the test is stopped before the time for full reaction, the results can be quite misleading. The investigators do note that dissolution of the octahedra is possible with strong acids and dissolution of the silica tetrahedra is possible with strong bases. Consistent with this expectation, Gipson (1985) found that permeation of bentonite–silty sand mixtures with acidic leachate high in calcium and other inorganics resulted in hydraulic conductivity increases with time. The naturally occurring clayey soils did not exhibit the same hydraulic conductivity increases.

Wu and Khera (1990) investigated the changes in the properties of a mixture consisting of a chemically treated bentonite and sand in a containment environment. They performed a series of tests to determine soil-chemical compatibility. Ballivy et al. (1992) discussed the effectiveness of injected cement grout under harsh environmental conditions. They described laboratory tests where the causes of grout degradation due to the effects of leachates have been investigated.

Another study (Peterson and Glendon, 1985) suggests that the buffering

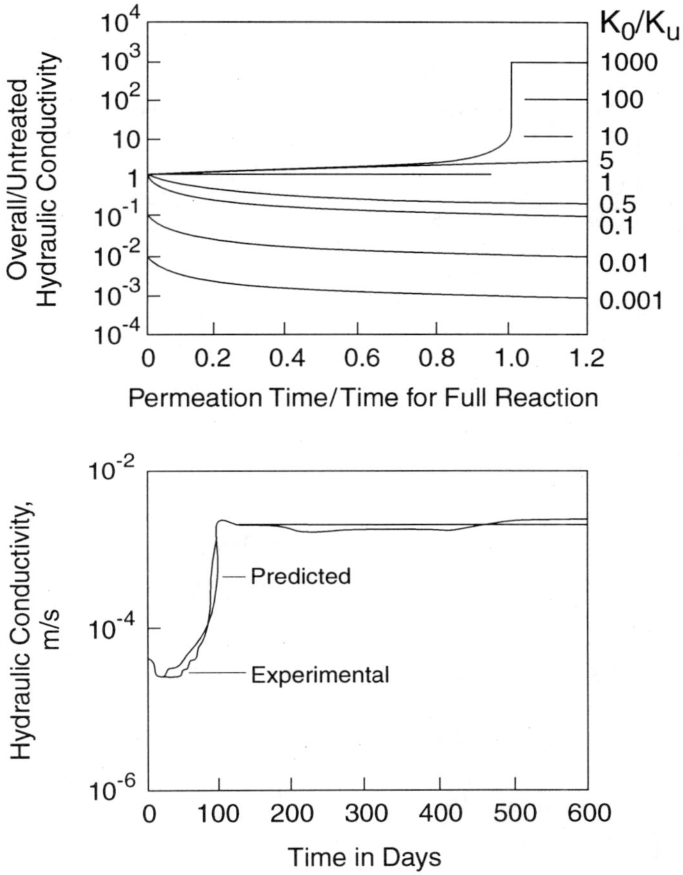

Figure 3.23 Impact of permeation time (Jefferis, 1992).

capacity of a soil may enable the permeated soil to continue to buffer acidic leachate for an extended time (in excess of 30 pore volumes). In a study testing soil–bentonite backfill, gradual and continual increases in hydraulic conductivity were observed for an acidic permeant (pH 1.0) as shown in Fig. 3.24 (Evans et al., 1985b).

While individual studies are useful in assessing the factors influencing the compatibility of clayey materials, it is also useful to consider literature reviews and syntheses of compatibility studies previously undertaken. In one such study, Mitchell and Madsen (1987) concluded the following:

1. Hydraulic conductivity changes can be understood from the perspective of the influence of chemicals on the soil fabric.
2. Chemical influences are likely to be greater with high water content ma-

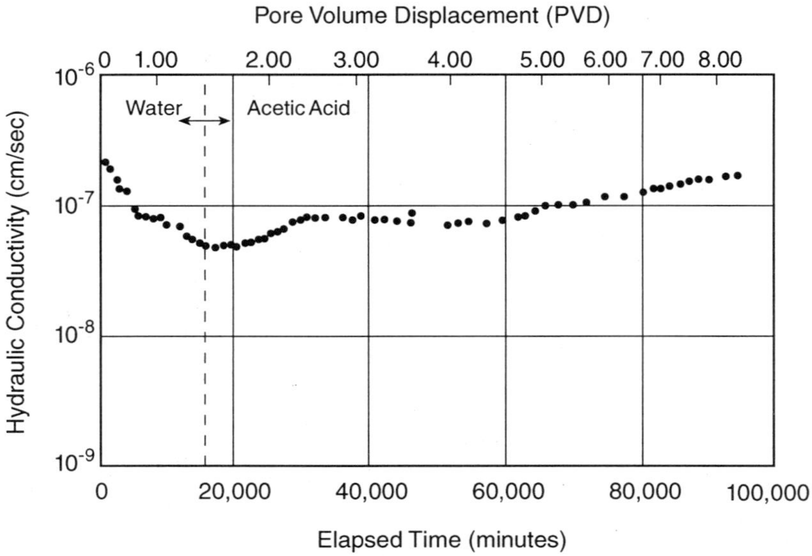

Figure 3.24 Permeation of soil–bentonite with acetic acid (Evans et al., 1985b).

terials (e.g., slurry wall backfill) than low water content materials (e.g., compacted clay).

3. Inorganic chemical effects are consistent with diffuse ion layer models.
4. Organic chemical effects are influenced by their dielectric constant, polarity, and concentration.
5. Test method selection may influence results.
6. In almost all cases, concentrated organics will cause shrinking, cracking and hydraulic conductivity increases, whereas, dilute solutions have essentially no effect.

A more recent review (Shackelford, 1994) essentially confirmed the findings presented by Mitchell and Madsen (1987). Shackelford (1993) concluded that significant increases in the hydraulic conductivity of clayey soils can result from

1. Flocculation of the clay particles due to interactions with electrolyte solutions.
2. Shrinkage of the soil in the presence of concentrated organic solvents.
3. Acid–base dissolution of the soil.

This review also revealed that there was considerable evidence supporting the use of the Gouy–Chapman theory in understanding the changes in hydraulic conductivity of clay soils. It was also noted that, for nonpolar, hydrophobic organic permeants, large entry pressures are required to force the permeant into the soil pores. Since field conditions do not generally result in such pressures, the interpretation of laboratory test results in the context of predicting field performance may be questionable in some cases. Nonpolar liquids tend to fill and flow only through the larger pores.

Based upon the literature described above, a number of general conclusions regarding soil–bentonite compatibility can be inferred. These conclusions are

1. Soil–bentonite backfill permeated with concentrated organic fluids is likely to exhibit increases in hydraulic conductivity as compared to permeation with water.
2. Increases in hydraulic conductivity due to permeation with concentrated organic fluids or with inorganically contaminated liquids are limited; that is, the hydraulic conductivity initially increases and then levels off to a new equilibrium value.
3. Increases in hydraulic conductivity can be limited by limiting the interactions between the soil and the contaminant. The more non-colloidal material (gravel, sand, and silt) in the soil (while still maintaining the desired hydraulic conductivity) the less will be the magnitude of hydraulic conductivity increase.
4. Permeation of soil–bentonite backfill with strong acids or bases may cause dissolutioning of the soil skeleton causing increases in hydraulic conductivity, which continue as long as the soil is exposed to fresh acids or bases.
5. Soils have a buffering capacity that may delay the increase of hydraulic conductivity when permeated by acids or bases.

Based on the available data it appears that the probability for incompatibility between the subsurface contaminants and the vertical barrier materials is low. The probability for degradation in cutoff walls increases in zones or portions of the wall where the organic constituents are in NAPL form. In this scenario, DNAPL could be found at the base of an aquifer or LNAPL could be found floating on the groundwater and, in both cases, the NAPL could be in contact with the cutoff wall. In these localized areas (where NAPL is in direct contact with the cutoff wall) some degradation is possible. An anticipated inward gradient typical of most pump-and-treat options would act to mitigate the potential for degradation under these conditions and for escape of pollutants if degradation did occur. Where the wall is not in contact with NAPL, the results of the published literature indicate compatibility problems would not be expected. If no NAPL is anticipated, the probability for incompatibility can be expected to be quite low.

Laboratory investigation programs can be designed to provide site-specific evidence of the compatibility or incompatibility between vertical barrier materials and contaminants in the vicinity of the cutoff wall. Concerns regarding incompatibility would be alleviated for any given project, should the findings from these tests be negative.

3.6 SOIL–BENTONITE BACKFILL DESIGN

In order of decreasing importance for environmental applications, the desired characteristics for a soil–bentonite backfill include:

1. Chemical compatibility
2. Low permeability
3. Low compressibility
4. Moderate strength

The most important soil parameters which affect these characteristics are grain size distribution and the water content of the backfill. Soil–bentonite backfills have been successfully created from materials varying from clean sand to highly plastic clay. However, for the containment of hazardous waste, backfill requirements are necessarily more stringent than those used for conventional barrier applications. The difference in requirements results from two fundamental differences in expected performance. First, conventional dewatering applications may not require the same degree of "perfection" as these systems are temporary. Small leaks may be inconsequential from a dewatering standpoint whereas from a hazardous waste containment standpoint they may be significant. Second, since these dewatering applications are short term, considerations of long-term changes in the hydraulic conductivity in the cutoff wall are not necessary. For hazardous waste containment applications, the long-term permeation of the cutoff wall with contaminants may alter the permeability of the material. Thus, compatibility is very important.

For waste containment applications the backfill must be designed to minimize any potential changes in hydraulic conductivity. Generally, a lower hydraulic conductivity is required for hazardous waste containment applications than for conventional dewatering applications. These goals of long-term stability coupled with low hydraulic conductivity can be best achieved by fabricating the backfill from a well-graded soil blended with soil–bentonite slurry. This well-graded soil should contain all particle sizes, including coarse, medium, and fine gravel; coarse, medium and fine sand; silt; and clay. A recommended particle size range is shown in Fig. 3.25. Since the coarser granular materials in a well-graded backfill material are in point-to-point contact, a relatively low compressibility results. Furthermore, with the well-graded nature of the material, the pore sizes (which become progressively finer) are filled

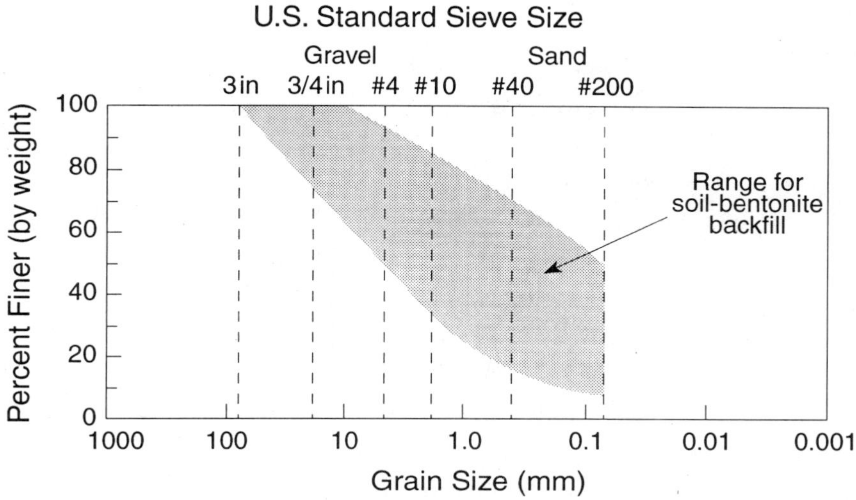

Figure 3.25 Grain size distribution for soil used in soil–bentonite backfill (Evans, 1991).

with progressively finer material (silts and clays) resulting in a material having a relatively low hydraulic conductivity. Finally, by using a well-graded material, the potential for long-term changes in hydraulic conductivity due to soil contaminant interactions is minimized. The components of the well-graded soil are relatively stable (sand, silt and low activity clay). Clay of low activity (e.g., kaolinite) is not subject to dramatic increases in hydraulic conductivity (Acar et al., 1985) as are those of high activity, such as bentonite (Evans et al., 1985a). This approach differs from simply adding more and more bentonite to a poorly graded soil to achieve a low hydraulic conductivity backfill. The swelling of the bentonite is potentially reversible in the presence of contaminants (resulting in hydraulic conductivity increases over the long term). It must be cautioned that a gap-graded material could possibly segregate during placement, and therefore should be avoided.

Therefore, the use of a well-graded backfill coupled with the limited addition of bentonite to further reduce the hydraulic conductivity is recommended. The reversible nature of the swelling of the bentonite may result in hydraulic conductivity increases in the presence of contaminants. As a worst case for hazardous waste containment applications, the hydraulic conductivity of the backfill may increase as compared to the hydraulic conductivity of the soil without the bentonite. By selecting a well-graded material, a relatively low hydraulic conductivity is found in the base soil, similar to the relatively low hydraulic conductivity of naturally occurring, well-graded glacial tills. The addition of bentonite–water slurry contributes to further reduce the hydraulic conductivity. In addition, the slurry provides moisture and enhances the fluidity of the backfill to allow it to displace the bentonite–water slurry used in

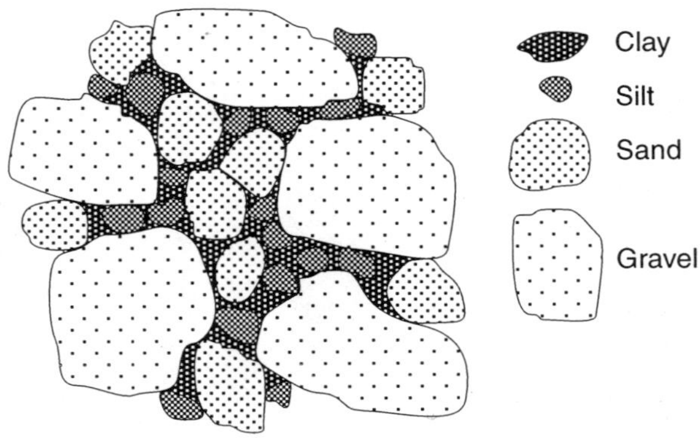

Figure 3.26 Schematic of well-graded soil.

the trench for stability. At the time of construction, the backfill must flow freely into the trench and displace the bentonite–water slurry used to maintain trench stability.

Some designers consider filter cake a significant factor in attaining low-hydraulic conductivity of the completed soil–bentonite slurry cutoff wall. However, for environmental applications, the added benefit of the filter cake should be considered only as an additional safety factor. This conservative approach is recommended for several reasons. First, the integrity of the filter cake cannot be ensured after backfill placement. As the backfill slides into the trench, the backfill may scrape all or part of the filter cake from the trench wall. Second, the filter cake consists of high-swelling bentonite. It is well documented that bentonite swelling is reversible in the presence of many contaminants (Alther et al., 1985). Thus, it is recommended that any filter cake contribution to the low hydraulic conductivity nature of the wall be considered simply as an additional safety factor.

Backfill is typically mixed along the side of the trench using materials excavated from the trench and then blended with bentonite–water slurry. For additional control over mixing, a remote batch process can be utilized. The mixed backfill is transported by truck to the trench for placement. For waste containment applications, this is the recommended method of mixing. For hazardous waste containment applications, pumping of backfill is not recommended because of the high water content of the mixture.

It is concluded that a well-graded soil (consisting of a blend of gravel, sand, silt, and clay) blended with bentonite–water slurry (resulting in a bentonite content of 1% in the mix) can result in a backfill of low-hydraulic conductivity, low compressibility, and greater resistance to degradation by contaminants than a backfill containing a very high percentage of fines in the mixture. The recommended design backfill is shown schematically in Fig. 3.26. The finer particles plug progressively finer voids, leaving only the smallest voids to be filled with the clayey fines and bentonite. A natural analogy is glacial till, which is generally well graded and possesses a low hydraulic conductivity.

3.7 DOCUMENTATION OF PERFORMANCE

Few data (published or otherwise) are available regarding the performance of vertical barriers constructed as part of waste containment systems. This is a result of two factors: (1) the means by which to ascertain the performance of the entire system are expensive and complex; and (2) corrective measures are difficult, at best, and consequently there is little motivation to measure the performance of the completed system. At present, the integrity of vertical barriers is based on construction quality control testing and, occasionally, postconstruction sampling and laboratory testing. Nondestructive testing offers some hope for improved confidence in the completed cutoff wall.

3.8 CONSTRUCTION QUALITY CONTROL

Construction quality control (CQC) procedures used for vertical barriers constructed to limit the lateral migration of chemicals in groundwater are described in this section. The barrier technologies considered apply to slurry walls (both soil-bentonite and cement–bentonite) and various grouting methods (permeation and jet grouting) since these techniques are the most commonly used for containment. Other techniques, such as vertical geomembranes, deep soil mixing, and sheet piling, have been addressed from a technical perspective, but CQC issues have not been widely discussed in the literature, probably because these methods have had limited use in containment applications.

3.8.1 Soil–Bentonite Cutoff Walls

a. Preconstruction. A CQC plan should be developed as part of the design and should reflect the technical aspects of the design (Barvenik, 1985, 1986). One purpose of the CQC plan is to identify substandard construction and the CQC plan should also identify remedial techniques to be implemented in the event that such problems arise. The CQC testing should be sufficiently inexpensive to allow a statistically significant number of tests to be conducted and should utilize simple equipment to facilitate field use. The CQC tests should be expedient thereby permitting identified defects to be remedied, yet sufficiently accurate to provide a meaningful expectation of postconstruction performance. The plan should reflect the two potential modes of failure, namely, a wall material with nonconforming permeability and one having discontinuities.

Planning involves identification of a suitable borrow quantity that will be mixed with bentonite-water slurry for the soil–bentonite backfill. Some projects use spoil excavated from the slurry trench for backfill, and others import borrow. The decision relies on such issues as the uniformity of the proposed soil borrow, its gradation, and the extent of chemical contamination of the excavation spoil. The need for compatibility testing is identified at this time. For example, if the wall is designed to have an inward gradient, the necessity for intense compatibility testing of the proposed backfill mixes is reduced as compared to the case where groundwater manipulation is not part of the design.

Another design component that relates to CQC is the quality of available water for use in mixing the bentonite slurry. The water quality affects the ability of the bentonite to hydrate, which is one of the most important characteristics (Xanthakos, 1994). Water, other than that from a public water supply system, should be evaluated for suitability.

The CQC plan should contain provisions for testing of the bentonite using various tests as described by the API (1990). Compatibility testing may be done by mixing a laboratory soil–bentonite sample and permeating it with leachate or contaminated groundwater and monitoring the permeability changes with time. Schulze et al. (1984) estimate that a two-to-three pore volume replacement is required to affect changes in the slurry wall backfill properties.

b. Construction. Construction quality control testing during construction consists of three phases: (1) testing the bentonite slurry, (2) assuring continuity of the wall, and (3) testing the soil–bentonite backfill. Each phase is discussed separately in this section.

Slurry Testing. The slurry consists of a mixture of bentonite and water and serves several important functions in the construction process. It supports the excavation, preventing soil from sloughing into the trench, and consequently is an important factor from a wall continuity and trench stability perspective. The slurry forms a filter cake along the trench side walls, limiting the amount of slurry that might seep from the trench. It also suspends excavation spoil, preventing it from collecting at the bottom of the trench and carries cuttings in the slurry volume (Xanthakos, 1994).

Bentonite is mixed with water that has been determined to have acceptable properties. After being mixed, but before being placed in the trench, the slurry is sampled and tested. Common tests are viscosity, unit weight, and filtrate loss (however, pH, sand content, and gel strength may also be tested). The viscosity may be measured using a Marsh funnel, the unit weight with a mud balance, and the filtrate loss with a filter press. These tests and associated equipment are described elsewhere (API, 1990).

When the slurry test data are satisfactory, the slurry is pumped into the trench. Once in the trench, the slurry is often sampled from the bottom of the trench using an appropriate device to obtain a representative sample which must therefore be capable of trapping slurry from the bottom of the trench. The sample is tested for unit weight, which must remain sufficiently below the unit weight of the backfill for proper displacement by the backfill (USEPA, 1987). Slurry from the trench can be tested for filtrate loss to indicate any adverse chemical effects from the soil on the slurry. In this case, the results from the filtrate loss test will be greater than those measured on the slurry before it was introduced into the trench. The slurry in the trench should also be tested for viscosity.

Trench Testing. This aspect of CQC testing is important from a wall continuity perspective. The CQC aspects of the excavation include verticality, continuity, depth of penetration into the aquiclude, bottom cleaning, and stability. The trench verticality is assessed by measuring the verticality of the excavation equipment with a level on the backhoe boom or tracks. Continuity is assessed by sweeping the trench with the backhoe bucket. It also can be checked by dragging a vertical probe along the trench bottom and observing blockages.

The penetration of the trench into a natural aquiclude has a significant effect on the performance of the wall. In fact, Grube (1992a) states that this aspect requires intensive quality assurance. The CQC personnel should measure the depth of the trench using a rigid probe when the backhoe operator indicates that the aquiclude has been reached. The excavation spoil should also be observed to confirm that the aquiclude has been reached and that construction of the key should begin. Following the key excavation, the depth should be measured again and the location of the measurement along the trench should be recorded.

Stability of the trench affects the continuity of the wall and should be monitored. Tension cracks in the ground surface parallel to the trench will precede a trench wall failure. These can result from the slurry level in the trench being less than several feet above the groundwater level (USEPA, 1984). Millet and Perez (1981) state that a minimum 3–5-ft slurry head above the groundwater level is required for stability. Sloughing from below the slurry level cannot be seen, but can be detected. If sloughing occurs during excavation, the material will be removed by continued excavation of the trench. If the excavation has been completed, sloughing can be identified by frequent sounding of the trench before backfilling begins. Material identified can then be removed by excavation or bottom cleaning. Sloughing, which occurs during backfilling, can leave a window in the wall. This window can be monitored indirectly by sounding the profile of the backfill slope at frequent, regular intervals. Sloughed material will likely produce a mound on the backfill slope. If this is observed, the excavation equipment should be used to collect a sample at the suspected location for observation. If it is sloughed soil, it can then be excavated.

Backfill Testing. This phase of the CQC program involves measuring the properties of the soil–bentonite backfill placed into the trench. The following properties are measured: unit weight, slump, gradation, bentonite content, and permeability (USEPA, 1987). The unit weight should be monitored since it must be sufficiently greater than the slurry for proper displacement. The slurry unit weight should be at least 15 pcf less than the backfill (USEPA, 1984).

Slump testing is the same as that done for concrete (ASTM C143). Slump is measured to provide an indication of the ability of the backfill to flow in the trench. A low slump backfill will not flow as easily and will result in a steep backfill slope compared to backfill having a higher slump. The slump is measured before the mixed backfill is placed into the trench. Typically, slump is in the range of 4–6 in.

Backfill gradation may be the least critical property measured during construction because backfill proportioning should have been done during design. Its primary value is to measure the amount of fines in the backfill. This test compliments the permeability testing.

Bentonite content in the backfill can be estimated using the methylene blue titration test. This is an indirect test in that it measures the cation exchange potential of the backfill and must be calibrated against the backfill mix before the addition of bentonite. This test is used in conjunction with permeability tests, particularly if a high-permeability value is measured, to determine the amount of bentonite (since a lower than specified bentonite content may be the cause of the high permeability). This test has limited reliability at low bentonite contents (<1–2%).

The most direct measure of soil–bentonite backfill performance is the permeability test. The USEPA (1987) describes the advantages of using an API fixed ring to make permeability measurements on samples of the backfill mix. The results of tests done using this method should be compared to data from permeability tests on identical samples using triaxial equipment during design. This is done to verify the accuracy of the test method for each backfill design mix. This test method has been found to provide reliable data in a relatively short time compared to open system triaxial testing. Its simplicity and relatively short test time enable a large number of samples to be tested, compared to open system triaxial testing, making it well suited for production field situations since the tests can be performed in a field laboratory. Alternatively, closed-loop triaxial testing can yield test results in 24 hr; the same time required for API filter press tests.

3.8.2 Cement–Bentonite Cutoff Walls

The process of constructing a cement–bentonite cutoff wall differs in some ways from that used for a soil–bentonite wall. Therefore, the CQC program for a cement–bentonite wall differs from that of a soil–bentonite wall in some respects but is similar in others. The methods used to construct the cement–bentonite wall also affect the CQC plan.

A cement–bentonite wall can be constructed by using the cement–bentonite–water slurry to support the trench. This material sets up in the trench and becomes the cutoff material. Thus, in a cement–bentonite wall, no backfilling is required. Thus, the fundamental difference between a soil–bentonite wall and a cement–bentonite wall is that a cement–bentonite wall can utilize a cement–bentonite–water slurry to support the trench as well as to set and become the cutoff wall, thereby obviating the need to backfill the trench.

a. Preconstruction. The first step in monitoring construction of a cement–bentonite wall is to develop a site specific CQC plan that reflects the project conditions, as in the case of a soil–bentonite wall. A condition that is project specific is whether the wall will be constructed continuously or in panels. A wall constructed continuously may require retarders in the mix to delay the wall set.

The permeability of the wall may vary by project, and the permeability should be estimated by conducting laboratory tests on proposed design mixes of cement and bentonite. The permeability may vary with the water/cement ratio and the water/bentonite ratio, and is typically higher than that achievable with soil–bentonite.

b. Construction

Slurry Testing. The bentonite–water slurry used for a cement–bentonite wall initially is the same as that used for a soil–bentonite wall. The same properties are measured for this application; that is, unit weight, viscosity, filtrate loss, and pH. These properties should be measured before cement is added to the slurry because the bentonite slurry reacts with cement and the properties will change. Millet and Perez (1981) recommend that mixed bentonite slurry be allowed to fully hydrate before mixing with cement. Grube (1992a) states that this hydration time should be at least 1 day but is strongly dependent on the type of mixer.

Trench Testing. The continuity of the cement–bentonite wall will affect its performance in the same way that it affects the performance of a soil–bentonite wall. The same procedures need to be followed as described for the soil–bentonite wall to identify sloughing and the proper key depth. However, a cement–bentonite wall adds another requirement, because the wall sets up overnight. This occurs regardless of whether the wall is constructed in panels (without a retarder additive) or continuously (with retarders). Thus the wall must be properly overlapped between panels or sections in order to avoid higher permeability windows between panels. Trench sounding is required to identify the extent of the overlap. Since there is no backfilling with cement–bentonite walls that use a cement–bentonite slurry, there is no need to sample the slurry from the bottom of the trench for unit weight measurements.

Mix Testing. The cement–bentonite mix should be sampled and its properties measured and compared to the design parameters to check proportioning at the mixing plant. The cement–bentonite mix is tested for unit weight, water content, sand content, permeability, compressive strength, strain to failure, and shrinkage potential (Barvenik, 1985, 1986). Millet and Perez (1981) also recommend that the sand content of the cement–bentonite slurry be measured.

3.8.3 Permeation Grouting

Permeation grouting is a general term, which refers to the construction of a vertical barrier by pumping cement grout into the ground to fill voids within the geologic matrix. The purpose of the grout is to fill voids in the underground matrix that would otherwise be conduits for groundwater flow, thereby reducing the permeability of the matrix. The matrix may be either soil or rock, and the voids could be rock fractures or pores between the soil grains. A series of grout holes along a line having at least three rows in parallel can be used to form a grout curtain.

Grouting projects usually rely on observations during grouting in order to make necessary adjustments. Thus the program requires a design–grouting–observation–redesign–regrouting cycle (Einstein and Barvenik, 1975). The most common grout is cement. However, the grout type should be selected based on the size of the voids to be filled (Welsh, 1992).

a. Preconstruction. Several factors are considered before construction begins and must be taken into account in developing the CQC plan. The compatibility of the grout-soil mixture should be studied if the barrier will be subjected to permeation with chemical contaminants. The permeability of the media and the size of the voids that will be grouted affect the initial grout viscosity, that is, the water/cement ratio (and the need for fillers), grout pressures, and the hole spacing that will be used. The required width of the barrier will dictate the grout hole array.

b. Construction. The CQC for grout barriers can be viewed from two perspectives, testing the grout material and testing the effectiveness of the grout in the formation. The quality of the grout mix can be checked by pouring a sample into a tray to a depth of about 1-in. and allowing it to cure overnight. The sample should be broken the next day and a vertical section should be examined for banding, an indication of poor mixing. The sample should be uniform (Houlsby, 1990).

A method to check the water/cement ratio of the mixed grout in the field is to fill a mud balance with grout in order to measure the specific gravity. The specific gravity can be related to the water/cement ratio using published tables (Houlsby, 1990). This is a quick test that provides a measure of how well the grout is mixed. Another indirect method of measuring the water/cement ratio is using a flow cone, although this is considered less reliable than a mud balance. One type of cone is the Marsh funnel used for drilling mud (bentonite slurry). There are other types of flow cones but there are no established standards for grout.

According to Einstein and Barvenik (1975), there are two basic types of control for grout barrier construction: primary and secondary. The three types of primary control are referred to as permeability, piezometer, and coring. The secondary controls are pressure and grout take (i.e., the volume of grout injected into a hole).

For permeability control, a section of the wall is constructed and a hole is made within the barrier at the farthest distance from the grout hole. This serves as the pumping hole. Observation holes are drilled outside the limits of the barrier. Water is pumped from the pumping hole at a constant rate while monitoring the observation holes. The observed response can be related to the permeability. Alternatively, a water pressure test, which requires no observation wells, can be performed. Water is pumped into the pumping hole and the pressure and rate of water pumped can be related to permeability if the geometry and dimensions of the grouted zone are known.

Another primary control method is piezometer control. This method involves installing piezometers upstream and downstream of the grout barrier and monitoring the water levels as the barrier is constructed. A difference in the water levels is an indication of the barrier effectiveness.

The third primary control method is the coring method. This technique involves drilling into the grout barrier and collecting samples for observation. The samples are examined to estimate the amount of grout that has penetrated the medium and the effectiveness of the barrier.

Secondary controls are relied upon once a procedure is selected that provides acceptable results. This method of control assumes that grouting pressures and volumes can indicate successful grouting. In pressure control, it is assumed that pumping grout at a certain pressure will force a given volume of grout into the medium to a given dimension. If additional pressure is required for that volume of grout, this indicates a change in the subsurface conditions and the procedure needs reevaluation.

The concept of grout take control is similar to pressure control. If the grout take differs for a given pressure, the procedure must be modified.

3.8.4 Jet Grouting

Jet grouting differs from permeation grouting in that rods are inserted into the ground and grout is pumped laterally through nozzles under very high pressure from the bottom of the rods as they rotate, mixing with the surrounding soil. The grout may also be pumped with air and water, depending on the system being used (Kauschinger et al., 1992). This process forms soil cement columns as the rods are withdrawn from the hole. The columns form a wall when they are placed in a line, and the continuity of the wall depends on the quality of the overlaps between the columns.

The equipment used for this technique is highly computerized enabling the operator of the equipment to closely monitor the process. The operator receives readouts of flow rates, quantities injected, and materials used as the column is being formed, enabling modifications to be made (Welsh and Burke, 1991).

Thus, this technique is similar to permeation grouting in that experience is used along with the data collected during construction so that the desired results are achieved.

Because the soil is disturbed (i.e., it is bulked) by the process and some is displaced by the grout, soil is ejected from the grout holes and must be disposed.

a. Preconstruction. The CQC program for jet grouted barriers is similar to that for a grout curtain by permeation grouting. Walker (1993) describes an approach in which soil samples are collected before construction from the area proposed for jet grouting, and these soils are mixed in the laboratory with various amounts of grout. The mixes are tested for permeability to estimate the grout percentage necessary to satisfy specifications for field performance.

Another preconstruction technique to estimate the size of the grout column created by jet grouting is to build a test jet grout column in the soil using the proposed grouting procedures. The test jet grout column is then excavated after cure and the dimensions of the column are measured (Grisham, 1992).

b. Construction. The CQC methods to measure the quality of a jet grouted cutoff wall are similar to those used for testing a permeation grout barrier. Hover (1993) describes a process that includes doing a pump test before construction to establish baseline conditions. Following construction of the barrier wall, holes are drilled at the overlap of the jet grout columns for in-place permeability tests. These could be variable head permeability tests using a packer. Additionally, piezometers are installed both upstream and downstream of the wall to evaluate the effectiveness, as described by Einstein and Barvenik (1975).

4 Bottom Barriers (Floors)

4.1 INTRODUCTION

The construction of a bottom barrier to control the vertical migration of leachate from a source downward into the soil and groundwater has been infrequently attempted at existing hazardous waste disposal sites. However, the option of using naturally occurring low-permeability geologic formations, such as a thick clay or sound rock, is a practical alternative. A cost-effective way of controlling leaching from a waste site is by the construction of a horizontal migration control (sidewall) barrier around the buried waste and securely keying its bottom into a natural low-permeability stratum, such as a low-permeability clay or rock formation located below the waste.

When no naturally occurring low-hydraulic conductivity stratum exists at reasonable depth beneath a waste site, consideration can be given to the construction of a vertical migration control (bottom) barrier. This can be accomplished several different ways, such as, by using grouting techniques, or employing a combination of tunneling, installation of geomembranes, and grout or some kind of slurry mix. Other less proven approaches include ground freezing to stabilize soils (Iskander, 1987) and the creation of electrochemical barriers to contaminant transport that can be installed across compacted clay liners (Yeung, 1990).

Bottom barriers have been built in the United States using grouting techniques for keeping water and contaminated water out of several sites (Burke, 1993). In Europe, consideration is being given to the construction of bottom barriers using a system of trenches, tunnels, and geomembranes (Jessberger, 1993b). In Russia, an integrated method of grouting saturated rock has been developed for installing vertical and inclined shafts and horizontal tunnels in connection with the Soviet coal-mining industry. Significant research efforts over the past two decades in a variety of disciplines have led to new and improved mine construction and operation in which grouting technology has been a key factor (Kipko et et., 1991).

The following sections describe various methods that might be used to construct a bottom barrier below a buried waste. The sections give details on how the barrier can be constructed and discuss the advantages and disadvantages of each method.

4.2 NATURAL BOTTOM BARRIERS

Possibly the best way to control the vertical migration of leachates is by constructing a vertical control barrier around the buried waste and keying it into an existing low-permeability geologic formation (clay or rock) below the waste.

If the bottom geologic formation is clay, or silty clay, it should have certain required properties. The desirable properties of clay liners and the factors that may affect their long-term stability are thoroughly discussed by Mitchell and Jaber (1990) and Daniel (1987). Similar arguments can be advanced for the desirable properties of natural clays that are being considered for bottom barriers. Most importantly, the clay must have a low permeability, be continuous, of sufficient thickness, and it should also have low diffusivity, as molecular diffusion can result in leachate migration across the barrier. The bottom clay should also have sufficient ductility to accommodate differential settlements and possible earthquake movements. Furthermore, it is required that the bottom barrier clay be reasonably homogeneous without pervious lenses and pockets, or larger rock fragments. Also, overconsolidated clays with fissures, that may have opened up when the original excavation for the site was made, are not as desirable for bottom barriers as are softer clays.

When a rock formation is being considered as the natural bottom barrier, it must be without joints, faults, fissures, cracks, or solution channels, and it must not be porous rock. As a rule, limestone, sandstone, and many igneous rocks are not suitable for bottom barriers, the latter because of potential shrinkage cracks. However, when faults and fissures in a rock formation are detected, such as by cross-hole pumping tests, these openings can sometimes be filled by grouting.

The above discussion illustrates that the problem may be complex. It is clear that each site will require a thorough analysis and it is ultimately the design engineer's responsibility to recommend the most appropriate containment approach. It seems reasonable to assume that in most field situations the saturated hydraulic conductivity of natural bottom barrier clays will be lower than 10^{-7} cm/s. However, such a low conductivity may not be a strict requirement if the natural clay barrier is thick, or if a sound rock underlies the clay.

The permeability of the clay must be determined in the laboratory on carefully obtained undisturbed samples, using a consolidometer-type permeameter and a confining pressure that is equal or less than the *in situ* pressures under the waste. The samples for testing should be obtained by random sampling. Daniel (1990) recommends the division of the area to be tested into a grid pattern, assigning a number to each grid, and using a random number generator to locate the grids where the samples should be taken.

Furthermore, the bottom clay barrier should be checked for chemical compatibility with the waste leachates. Specifically, the compatibility of the clay with the leachate in question should be tested in the laboratory using the consolidometer-type permeability apparatus with a confining stress that is again representative of the in-field conditions. Enough leachate should flow through the samples to determine the final permeability of the clay. This subject is well detailed and summarized in Mitchell and Madsen (1987), Mitchell and Jaber (1990), and Daniel (1987). Directionally-drilled wells, combined with sampling and analytical testing can be used for detection of chemical transport across the barrier. New *in situ* borehole monitoring sensors and on-site field instrumentation are presently being developed for this type of application (Robbat, 1993).

Since at most sites there will be a low-permeability bottom geologic formation at some depth, a decision will have to be made whether to go down into this formation with the barrier side walls, or to stop at a shallower depth and construct a bottom barrier. This decision is partly based on the practicality of reaching the low-permeability bottom formation, and certainly on the economics of the alternatives. Even if the naturally occurring low-permeability bottom can physically be reached, a cost analysis should still be made to find the more cost effective of the two approaches. The break-even depth will be a function of the unit cost of each (i.e., the sidewall barrier and the bottom barrier), the depth of the natural clay or rock, and the surface area of the waste site (see Section 4.5).

Consideration must also be given to the tightness and reliability that can be achieved in the construction of a bottom barrier and its connection to the vertical sidewall barrier. Therefore, aside from strict economic considerations, the estimated level of risk reduction should also be factored into the decision, and in some instances the more reliable natural bottom barrier may be preferable, even if this alternative is the more expensive option. Conversely, if the water tightness of the naturally occurring bottom clay or rock is questionable, then the engineer may resort to constructing a bottom barrier at depths near the natural barrier.

4.3 GROUTED BOTTOM BARRIERS

4.3.1 Grouts and Grouting

A discussion of grouts and grouting techniques for constructing vertical sidewall barriers is given in Section 3.3 and many of the details will not be repeated here. This section focuses on the description of grouts applicable to the construction of bottom barriers.

To place grout into the subsoil, the grouting tube must be driven or drilled into the ground, the grout must be prepared, and the grout pumped through the tube into the subsoil. The drilling of the borehole for the grouting tube can be vertical, horizontal, slant or multidirectional. The latter is called directional drilling.

In the past, grouts have been used to increase the shear strength of a soil and to decrease the permeability of a soil formation. Grouts have historically been used for horizontal migration control barriers (cutoff walls or grout curtains) in dam construction by actually forming a wall in a soil formation, or filling the fractures in a rock formation. Grouted sidewall barriers have also been used to contain buried wastes (Gazaway and Jasperse, 1992; Mitchell and van Court, 1992a). Customarily, such barriers have been tied into low-permeability soil formations located below the waste for complete encapsulation.

If there is no low-permeability soil or geologic formation located below the buried waste, or the low-permeability layer is too deep to reach with a grout curtain, directional drilling may be used in conjunction with grouting to form a migration barrier beneath a buried body of waste. The sidewall barrier and bottom barriers must be carefully constructed to form a watertight seal at their joint.

To obtain a low-permeability layer using grouting, the grout itself must have an initial permeability that is low enough to satisfy the cutoff requirements. For example, since the USEPA requires hydraulic conductivities for sanitary and secure chemical landfill liners to be equal to or lower than 10^{-7} cm/s, one would not use a grout that has an initial hydraulic conductivity greater than 10^{-7} cm/s. Not all grouts meet this criteria. For example, one must be careful of relying on the initial permeabilities of sodium silicate grouts because they may undergo a phenomenon known as syneresis upon setting (syneresis is the exuding of water from the grout, which causes shrinkage of the grout and the potential opening of voids).

Once the grout layer is in place, the design engineer must be concerned with conditions (such as the presence of particular chemicals) that may cause an increase in the permeability of the in-place grout. When looking at the effects of chemicals on a grout, one must consider the compatibility between the contained leachate and the grout. The chemicals in the leachate may increase the permeability of one grout but not that of another. Chemicals that are destructive to Portland Cement grouts, that is, which increase their permeability, include: sulfates, carbon dioxide, humous acids, and very soft water (Nonveiller, 1989). Sulfates deteriorate the calcium compounds in the grout, which can be prevented by using sulfate-resistant cements. Any free carbon dioxide above the balanced amount (gas or dissolved in the groundwater) dissolves the free lime by turning the insoluble calcium hydroxide into soluble calcium salts. This can be countered by using microsilica in the grout. Finally, very soft water can dissolve the calcite salts in cement grouts and damage their integrity. On the other hand, Portland Cement grout has shown remarkable resistance to a variety of chemicals, especially when mixed with microsilica. Exceptionally low permeabilities have been obtained and sustained with microfine cement and microsilica grouts (Bowers et al., 1988). Not all grouts react in the same way when exposed to particular chemicals. For example, the permeability of acrylate grout increases when it comes in contact with acetone, aniline, and methanol, while the permeability of urethane grout increases in the presence of sodium hydroxide (Bodocsi and Bowers, 1991). Consequently, a compatible chemical grout must be chosen for each individual site. At the time of grouting, destructive chemicals may already have leached into the groundwater from a hazardous waste landfill. To reduce the effect of these chemicals on the grout, actual site groundwater should be used when preparing the grout. If the grout sets or gels in the trial runs, then it probably will not be severely affected later.

One should also consider the pH of a soil when choosing a compatible grout. Portland Cement grouts perform best in neutral-to-basic soils (acidic soils can adversely affect the chemical reaction in the cement). The acceptable soil pH for chemical grouts also varies from grout to grout. If the soil pH is too high or too low, it can keep the chemical grout from gelling, or from forming a low-permeability barrier.

There are two main types of techniques that can be used in placing grouts: (1) injection permeation grouting and (2) jet grouting. Four types of drilling can be used for both permeation and jet grouting: (a) vertical drilling, (b) horizontal drilling, (c) inclined (slant) drilling, and (d) directional drilling. The different methods of grouting that may be used to construct a bottom barrier are discussed in subsequent sections of this chapter.

4.3.2 Construction of Bottom Barriers by Permeation Grouting

A bottom barrier or floor below a body of waste may be created by permeation grouting using either vertical, horizontal, directional, or slant drilling. Vertical drilling, which involves drilling through the waste, is generally not desirable. Generally, vertical drilling can be used to a maximum depth of 36 m, beyond which alignment of holes is difficult to control. The thickness of the grouted barrier must be great enough to ensure adequate watertightness. Field tests would have to be performed to determine the required thickness. In the tube-a-manchette method, the tubes would need to be filled with low-permeability grout after placement of the barrier had been completed.

Horizontal drilling techniques can be utilized by using trenches or grouting galleries, probably in conjunction with the tube-a-manchette method. Again, a field test should be performed to determine the required thickness. The horizontal drilling method would not involve drilling through the waste, but it may involve drilling through contaminated soils or contaminated groundwater below the waste. Additional costs associated with the horizontal drilling method include the excavation costs of the trenches or grouting galleries and the cost of dewatering if working below the water table. When employing the horizontal drilling method, working conditions would be more constrained than in the vertical drilling method, and operators may have to be protected against exposure to contaminants in the confined space. Dipping the holes slightly downward toward the center of the waste would minimize the flow of leachate out of "horizontal" drill holes.

Another method that could be used to contain buried waste is to form a low-permeability grout cone under the waste. Using inclined drilling, along with injection grouting (and tube-a-manchette), and targeting a common point below the waste, a low-permeability cone can be formed. The success of this method would depend on the accuracy of the drilling. Again, one row of grout

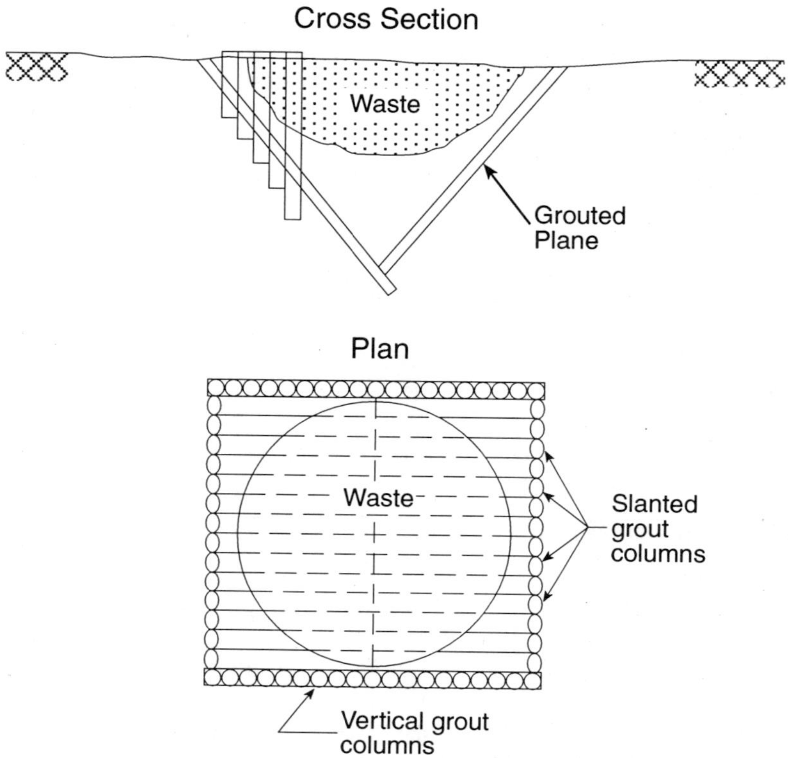

Figure 4.1 Bottom barrier formed by slanted grout columns.

columns would probably not provide sufficient watertightness, and a two- or three-row pattern may be necessary. Additional columns may have to be grouted to fill gaps in the cone. The accurate alignment of the grout holes is very important.

A variation of the grout cone involves the placement of two slanted planes formed by grouting under the buried waste. Each plane is drilled from opposite sides of the waste toward a common line running below the waste. The ungrouted ends of the containment structure would be created by two sidewall barriers (see Fig. 4.1). The grouted slanted plane barrier would probably be easier to construct than the grouted cone barrier, due to drilling inaccuracies and geometric layout problems that may be associated with the grouted cone.

As discussed in Chapter 3, the gradation of the native soil materials greatly influences the success of grouting, since many fine sand and silty soils cannot be penetrated. As discussed in Section 4.3.3, jet grouting does not have this limitation.

The approximate cost of injection grouting 1 yd^3 of soil ranges from \$35 to

175 using a cement or cement–bentonite grout, and from $150 to 350 using a chemical grout (1993 dollars).

The hydraulic conductivity of the grouted soil will vary greatly, from 10^{-6} cm/s, when pure Portland Cement grout is used, to 10^{-10} cm/s, when acrylate grout or microfine cement with microsilica grout is used. The conductivity of a soil grouted with sodium silicate will be relatively high, often in the range of 10^{-5} cm/s.

The ductility of the grouted soil will be moderately high with the acrylate and urethane grouts, and low with the cement-type grouts. However, cement-bentonite mixtures will make the grout more ductile, and also less permeable. A soil with sodium silicate grout will generally have relatively low ductility.

4.3.3 Construction of Bottom Barriers by Jet Grouting

Similar to injection (permeation) grouting, jet grouting can be used to create both bottom and vertical barriers. In fact, to create a watertight seal at the connection between the two types of migration control barriers, the use of jet grouting is preferred for both (Burke, 1993).

Jet grouting can be performed using either vertical, slant, horizontal, or directional drilling. Horizontal drilling for the purpose of creating a low-permeability horizontal bottom barrier would be limited to lengths of approximately 50 ft (Burke, 1993; Welsh, 1993), due to the lack of accuracy involved with horizontal drilling. Another limitation associated with horizontal jet grouting is due to the gravitational settling of the grout and the water. If horizontal jet grouting is to be used to place a barrier below a body of waste, trenches or grouting galleries would need to be excavated, adding significantly to the cost of jet grouting. However, new technology is being developed that combines directional drilling with a new configuration of jet grouting equipment, which may be useful to construct a slab-shaped grout barrier.

Vertical drilling combined with jet grouting may be used to create both vertical and bottom barriers to contain a hazardous waste site. The maximum drilling depth to create a low-permeability barrier by jet grouting is approximately 36 m (Burke, 1993; Welsh, 1993). Beyond this depth the lack of accuracy of the vertical alignment of the drilling cannot ensure complete filling of voids with grout between the columns. Although placing a bottom barrier beneath a body of waste by vertical drilling involves drilling directly through the waste, Welsh (1993) and Burke (1993) believe that with proper precautions this can be done without posing unnecessary risks to equipment or operators.

In order to construct a bottom barrier below a buried body of waste by vertical jet grouting, a column pattern like that illustrated in Fig. 4.2 would need to be employed. The ground surface would have to be leveled or, if possible, a level concrete pad would be poured to form a platform from which drilling accuracy could be assured.

Currently, contractors in the United States have the technology and ability

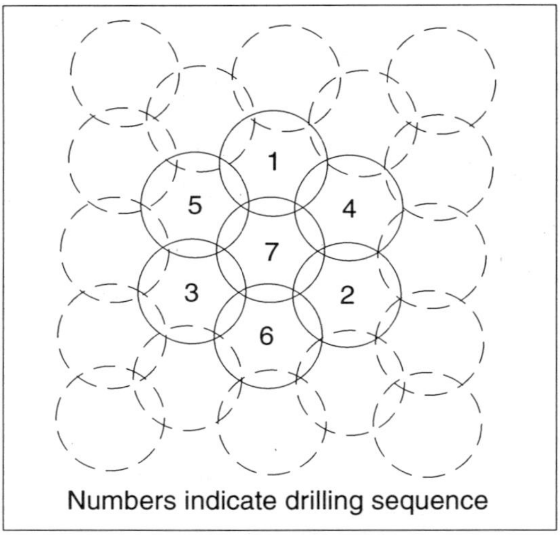

Figure 4.2 Geometric layout of jet grouted columns for bottom barrier.

to construct a jet grouted bottom barrier at relatively small sites. Jet grouted bottom barriers have been constructed at three sites in this manner; two sites in Florida and the other in New Jersey (Burke, 1993; Welsh 1993; Burke and Brill, 1993).

The construction of a low-permeability cone barrier or two low-permeability slanted plane barriers under a body of waste would be impractical using jet grouting since, in practice, jet grouted columns are not usually angled more than 30° from vertical. Therefore, the grouted columns used under a large site would have to be very deep in order for columns to meet at one point or along a common line, rendering the system uneconomical.

Overall, jet grouting should provide better barrier characteristics than permeation grouting. Jet grouted columns give a more consistent and defined cylindrical shape than the permeation grouted columns, which means that one can be more confident about the reliability of the low permeability of the barrier. Another advantage of jet grouting is that it can cut into an existing jet grouted column, thus ensuring that adjacent jet grouted columns form watertight seals. This technique is used in constructing both bottom and sidewall barriers. Jet grouting is also able to incorporate obstructions such as boulders and timber piles directly into the soilcrete mixture (Kauschinger et al., 1992).

A new sonic method has been developed to check the size of the cylindrical void in a grouting operation just before the introduction of the grout (Kauschinger, 1993). The soil is first removed by jetting with water. This is followed by the introduction of a sonic emitting and receiving device (that is attached to the tip of the drill rod), which measures the size of the water-filled cavity. The emitted sound waves easily travel through the water-filled cavity and are reflected from the soil sidewalls of the cavity. This measurement technique helps to ensure the proper placement of the grout.

The use of the triple-rod system of jet grouting is the most attractive because it can create a larger grouted column compared to the single- and double-rod systems and would, therefore, cover a specific area with fewer drill holes and grout columns. Also, the triple-rod system can create a pure grout-filled column with good quality control. However, the triple-rod system uses a generous amount of air which, if trapped, could cause the permeability of the columns to increase. It is recommended that at all sites grouted test columns be constructed to check on achievable column sizes, and that cores be obtained from the test columns and tested for permeability.

Bottom barriers can be constructed using both jet and permeation grouting techniques. In choosing between the two techniques, jet grouting would be the preferred method for placing a low-permeability bottom barrier below a buried body of waste for the following reasons: (1) good control of grout column dimensions; (2) good control of continuity of grout placement; (3) relative independence from heterogeneities in the soil; and (4) ability to use Portland Cement grout, which has high resistance to chemicals, yet is one of the least expensive types of grout. The Portland Cement grout, which cannot be injected in many soils, can be used with jet grouting in almost all types of soils. This latter consideration is a great advantage of jet grouting.

In summary, a typical jet grouted bottom barrier made with Portland Cement and bentonite may have a hydraulic conductivity as low as 10^{-7} cm/s. Research is underway to develop a mix that would further reduce this value, yet maintain good workability. The cost of constructing a bottom barrier by jet grouting ranges from \$50 to 350/yd^3 (1993 dollars). For example, the bottom water barrier installed near the port of Jacksonville, FL for a coal unloading structure cost \$90/ft^2 for the 7–8-ft thick mat (i.e., ~\$325/yd^3). The hydraulic conductivity of the constructed bottom barrier was approximately 5×10^{-7} cm/s. The strength of this bottom barrier was 800 psi, designed to resist uplift. Under a waste site, strength of the mat would not be a critical requirement so that the reduction of strength resulting from the addition of bentonite to the grout mix would likely be acceptable. Addition of bentonite to the grout mix results in low hydraulic conductivities (10^{-7} cm/s or less). Utilizing a high bentonite content produces a cement grout with low strength (20 psi), low hydraulic conductivity, and moderate ductility (Burke, 1993).

4.3.4 Construction of Bottom Barriers by Directional Drilling and Grouting

Directional drilling allows the drilling of a "parabola-shaped" path from one point on the ground surface to another point on the ground surface. Directional drilling can be used for the following purposes: (1) to allow injection of grout; (2) to facilitate the placement of drainage pipes below the ground surface; and

(3) to provide access for placing high technology sensors and other observation equipment below a constructed bottom barrier.

The key to the directional drilling method is the use of a steerable drill head. When the drill head is pushed, it changes direction because of the unbalanced soil reaction on the head. As a result, the drill head may be forced to go up or down as it is forced forward, giving the drill a curved path. Rotating the drill head will result in a straight drill path.

The drill rig is similar to that used in soil nailing or tie-back drilling. The drill is anchored to the surface and the drilling is started at some angle to the surface. The drill rig is hydraulically powered and is capable of performing both rotary and pushing motions. The drill rods are flexible and can bend to accommodate the shape of the drill path created.

The position of the drill head can be tracked using a radio transmitter that is located directly behind the drill head. The radio receiver is placed at the ground surface. The tracking is accurate to within 2–5 cm in any one direction, but is limited to depths of 8 m due to the inability of the radio waves to penetrate deeper (Murdoch, 1992a). The accuracy of the tracking method is affected by the presence of scrap metal (e.g., buried barrels) between the drill head and the radio receiver. New technologies (electronic accelerometers, electronic compasses, or inclinometers built into the drill head) are under development that may overcome these depth limitations.

If permeation grouting is used, the tube-a-manchette permits multiple injections. The drilling sequence would be the same as that outlined in the permeation and jet grouting sections. It may be necessary to use two sets or rows of grout holes on top of one another to insure watertightness. Also, a third row of holes in the very bottom may be used to detect discontinuities by reflection seismic tests (Witherspoon, 1993) and/or to monitor for any leachates that may seep through the barrier.

The minimum radius of curvature that directional drilling can achieve is on the order of 15–30 m. The maximum diameter hole that can be drilled is 1 m, using an expander that is pulled back through the hole after drilling (McMahon, 1993). The maximum length that can be drilled is between 600 and 750 m. Drilling rates of 300 m per 6 h have been achieved. At the present time, typical estimates for directional drilling with casing range between \$50 and \$100 per foot.

Directional drilling can be used in conjunction with both permeation (injection) grouting and jet grouting. If permeation grouting is used, a tube-a-manchette should be employed so that multiple injections are possible. The drilling sequence would be the same as that outlined in the permeation and jet grouting sections. Again, it may be necessary to use two sets or rows of grout holes to ensure watertightness and possibly a third row of holes to detect discontinuities employing reflection seismic tests (Witherspoon, 1993), and/or to monitor for leachates that may seep through the barrier.

If directional drilling with jet grouting is used, a special jet grouting tool may be employed to cut a rectangular swath in the soil below the waste and form a slab-shaped grouted body. This technique would create a continuous barrier under the waste and could encapsulate a waste site as wide as 300 m. Another proposed method of grouting for a low-permeability barrier by directional drilling uses a soil–grout mixer that may be pulled between two previously drilled adjacent holes, and, as it is pulled, it mixes the soil in front of it with a cement–bentonite grout. This procedure is repeated side-by-side until a continuous barrier under the waste is formed.

Directional drilling has been incorporated into a horizontal barrier installation system being developed by Brown & Root Environmental of Houston, TX (Dougherty, 1993). Brown & Root's developmental horizontal barrier installation system utilizes spoiless horizontal drilling technology to create small diameter horizontal bore holes with the drill pipe remaining in the uncased hole. Advantages to small diameter, spoiless, bore holes are (1) eliminates soil, mud, and water spoils resulting from drilling processes; (2) minimizes size and power required to place holes. Holes are placed in roughly parallel paths 10–20 ft apart (within 1 ft of the targeted path) in "parabola-shaped" paths under the site that is to receive the bottom barrier. The drill pipes left in the bore holes become guides for the subsurface tool bar that places the soil–grout barrier. The soil–grout placement procedure is repeated side-by-side through adjacent parallel bore holes until a continuous barrier under the waste is formed.

The main advantages of placing a low-permeability bottom barrier by directional drilling and grouting under a body of waste are (1) the waste would not have to be drilled through, as in vertical drilling; (2) no excavation would be necessary for trenches or grouting galleries like those needed in horizontal drilling; (3) the location of the individual holes can be tracked accurately, ensuring complete coverage of an area; (4) only one row of holes may be necessary to obtain low permeability if the "cutter–mixer method" or the special jet grouting tool is used; and (5) the size of the planar area of the body of waste is not as limiting a factor as in horizontal drilling.

One disadvantage of the directional drilling and grouting method is the limitation of the radio tracking system. Successful development of new technologies using accelerometers or inclinometers should greatly improve tracking. No known horizontal barriers have been constructed using directional drilling at this time, but several are in the planning stage for U.S. sites (Kauschinger, 1993; Murdoch, 1992a).

Karlsson (1993) recently described innovations in drilling and completion technology for horizontal wellbores where the wellhead is located outside of the contaminated area. Lehman (1993) described the "SUPERJET" method that uses an extremely high-pressure water jet to create large underground cylindrical cavities over 5 m in diameter, which can serve as one element in the construction of a horizontal barrier.

Materials used in forming horizontal soil–grout barriers result in grout similar to those used in classical jet grouting. Hydraulic conductivities of placed horizontal barriers are typically 10^{-7} cm/s with strengths ranging from 20 to

1000 psi. The body of knowledge and experience gained forming vertical barriers of cement, bentonite, and fly ash is directly applicable to grouts for forming horizontal barriers.

Although placement of horizontal barriers under waste sites has been demonstrated, it is an emerging technology for isolating or containing subsurface hazardous waste.

4.3.5 Construction of Bottom Barriers Utilizing Hydrofracturing and Grouting

Hydrofracturing is the intentional fracturing of a soil formation using water (or air, if pneumatic fracturing) pumped into the ground under high pressure. The first step in constructing a bottom barrier by hydrofracture is to drill a hole into the ground to the desired depth of the hydrofracture. A small horizontal slot at the bottom of the hole is cut using a jet, giving the fracture a place to start. This is followed by pumping grout (cement–bentonite or chemical) into the borehole under pressure. The pressure and volume of the grout are increased until the pressure suddenly decreases and the grout uptake increases, indicating hydrofracture has occurred. The grout pumping is continued, which forces the soil fracture to propagate and be filled with grout. The grout in the fracture forces the fracture to remain open. Upon the setting of the grout, the fracture attains low permeability.

Fractures typically develop in thicknesses of approximately 0.2–0.4 in. They often radiate uncontrollably from the borehole in a preferred direction and have maximum in-plane dimensions of 20–35 ft. In overconsolidated soils the fractures form parallel to the ground surface, and then gently climb toward the surface (and may break the surface on occasion). In normally consolidated soils the fractures are generally vertical and tend to propagate toward the ground surface. In overconsolidated soils the fractures can be stacked on top of one another at distances of 6–12 in. (Murdoch, 1992b).

Very sensitive tiltmeters installed at the ground surface above the hydrofractured area can be used to monitor the lateral extent of the hydrofractured zone (Witherspoon, 1993). Due to the small thickness of the fractures, the difficulty in monitoring their horizontal dimensions, and the operator's inability to control their direction, hydrofracture may not be a suitable choice for constructing a bottom barrier. Because horizontal hydrofracturing can only be performed effectively in overconsolidated soils, the sites where it can be used are limited. By making several hydrofractures on top of one another and drilling a dense pattern of holes, one could only reduce the flow of leachates, at best.

4.3.6 Construction of Bottom Barriers Using the Block Displacement Method

The block displacement method, or BDM (Brunsing, 1983) represents a variation of the hydrofracture method. Using BDM, a series of boreholes are in-

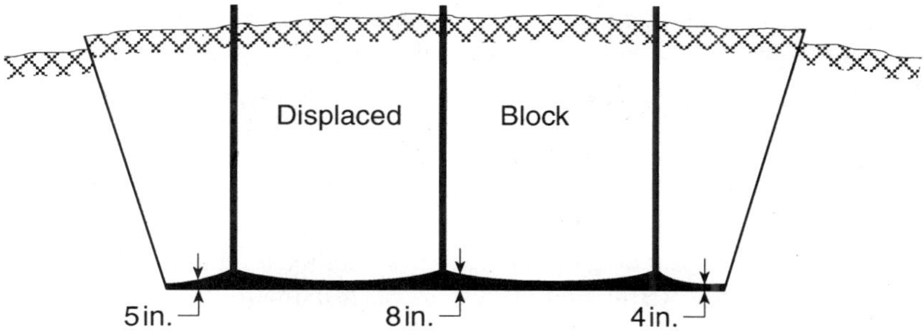

Figure 4.3 Cross section of displaced block.

jected simultaneously in order to lift a block (or mass) of waste and surrounding soil and to place a low-permeability bottom barrier beneath the block (Fig. 4.3).

The BDM requires drilling several boreholes around the perimeter of the waste, as well as through the waste. Once drilled, the perimeter holes are notched vertically with an air jet from bottom to top on two sides, so that the notches extend toward each adjacent perimeter hole (Fig. 4.4). The perimeter holes and their notches are intended to initiate shear planes, which permit the block to slide upwards. Next, several ''interior'' injection holes are drilled inside the perimeter to a depth that is below the buried waste. The interior holes are cased with a PVC pipe and notched in a circular pattern at their bottom using an air jet. Bentonite grout is pumped at high pressure into each interior injection hole simultaneously so that horizontal fractures are developed. The bentonite grout fills the fractures, forcing them to open wider and

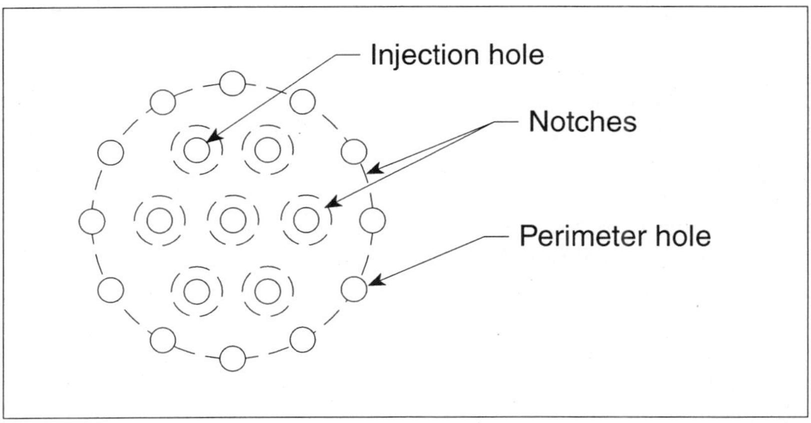

Figure 4.4 Plan view of injection and perimeter drill holes with notches to facilitate block displacement.

expand laterally. The thickening and expansion of the fractures force the block to displace upward. The bentonite grout also keeps the fractures from closing. The grout filled fractures act as a bottom barrier. Grout pumping is continued for several days at a slow rate in order to continue lifting the block and to ensure complete lateral coverage and filling of the fractures. It is postulated, that by using a sidewall barrier around the perimeter in conjunction with the BDM, complete containment of the buried waste can be achieved.

A BDM test was actually conducted in Whitehouse, FL by Foster-Miller, Inc. of Waltham, MA (Brunsing, 1983). A 60-ft diameter block was lifted and a low-permeability barrier was placed below the block at a depth of 23 ft below the ground surface. This was accomplished using 32 perimeter holes and 7 interior injection holes. After lifting, exploratory borings were drilled to determine the consistency and thickness of the bottom seal. It was found that the thickness ranged between 3 and 12 in. inside the originally planned perimeter, with the exception of one small area that was not successfully sealed and, in fact, was not lifted at all. In the area of zero lift, trenches were excavated to determine the cause, revealing that tap roots up to 12 in. in diameter ran through this part of the soil and were determined to be the cause of zero lift and lack of success in placing a barrier at this location. The actual effectiveness of the barrier was not determined at the Whitehouse, FL site.

In general, BDM would be limited in practical applications. The ability to place a continuous and effective barrier under a waste site cannot be done confidently using BDM. First, if the soil has large void spaces, excessive leak-off of the bentonite grout could occur, preventing the block from lifting and the formation of the horizontal seal. Second, the existence of roots, boulders, underground utilities, and other discontinuities present in the soil would also prevent the block from lifting. Finally, the size of the area that could be lifted is unknown, but it may be quite limited.

4.3.7 Sheet Pile Cutoff Combined with Injection Grouting

A buried waste may also be contained utilizing a combination of steel sheet piles and permeation grouting (Gerber, 1992). The steel sheet piles are driven at angles toward a common point (vertex) or line, located below the buried waste, forming a cone or wedge. Once the sheet piles are in place, the soil mass between the sheet pile wedge and the waste is filled with grout, using permeation grouting through multiple injection holes (Fig. 4.5). The grout fills the voids in the soil and the joints between the individual sheet piles, preventing further leachate migration.

The sheet pile and grouting method has been used to contain a waste site (Gerber, 1992) with the project being considered a success. However, the depth of the site that can be contained by this method is limited by the drivable lengths of the sheet piles (<100 ft). Also, it is difficult to drive inclined sheet piles and maintain alignment.

Another method that may be used to contain leaking underground tanks or

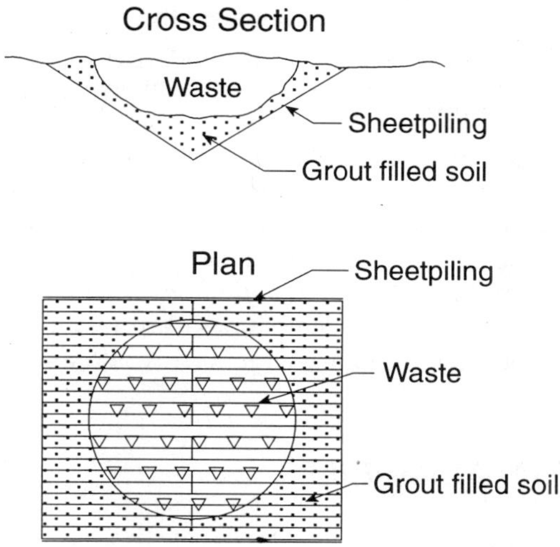

Figure 4.5 Cross section and plan view of bottom barrier formed by sheetpiling and grouting.

waste involves encircling the buried tank, or waste, with steel sheetpiling. Once the cylindrical sheet pile arrangement is in place, three clean access shafts (outside the contaminated area) are excavated around the cylinder. From these shafts or grouting galleries, horizontal holes are drilled and injected with grout. The horizontal holes are drilled in a radial pattern. The resulting grouted sheets form a low-permeability bottom barrier below the buried tank (Fig. 4.6). Once

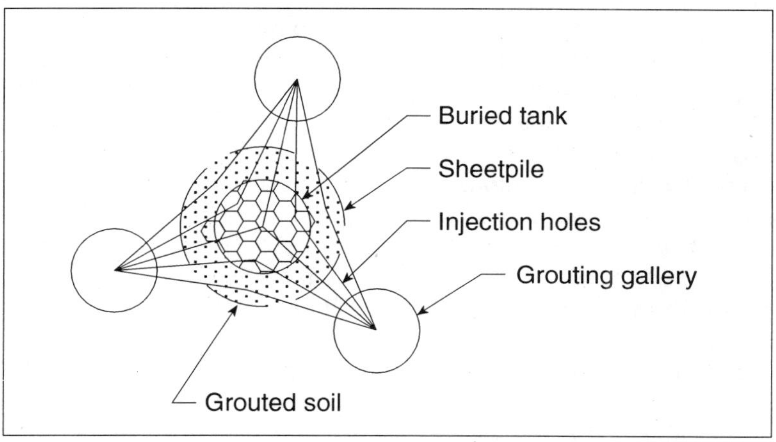

Figure 4.6 Plan view of vault containment formed by vertical sheetpiling and grouting.

the bottom barrier is in place, the soil between the buried tank and the sheet pile cylinder is filled with grout, using permeation grouting. The grout fills the voids in the soil and the joints between the individual sheet piles, preventing further leakage.

The formation of a cylindrical sheet pile sidewall perimeter coupled with grouting has been successfully used to contain leakage from buried storage tanks (Gerber, 1992) at the U. S. DOE Hanford facility. The method could also be used to contain other buried wastes. However, it would probably be limited to small sites, because of the limitation on horizontal drilling distances and the prohibitive cost of filling large volumes of soil with grout. The number of grouting galleries needed to grout a large area would also be expensive.

4.4 OTHER BOTTOM BARRIERS

4.4.1 The Zublin System

The proposed Zublin system results in the complete encapsulation of a waste site (Hausner et al., 1991; Jessberger, 1993b). The system results in a constructed barrier having a bathtub shape. Figure. 4.7a and b depict a typical plan view and cross section of the containment system during construction. The construction of the seal begins by excavating two parallel open trenches on two opposite sides of the waste site. From one of the trenches, large steel tubes are pushed and simultaneously excavated under the waste to reach the other trench. These tubes have a nominal diameter of between 7.5 and 9 ft, are parallel, and spaced 30 to 60 ft apart. They are also accessible to construction personnel. It is estimated that these tubes can be pushed to a distance of somewhere between 225 and 450 ft. The alignment of the tubes has to be carefully controlled.

A special excavator, called the ''sword'' (shown in Fig. 4.8) is the heart of the system. It is used to excavate (by jetting) an approximately 4.5-ft thick layer of the natural soil, keep the space open, and assist in the installation of a liner system between adjacent tubes (see Fig. 4.9). The liner system consists of a high-density polyethylene, surrounded by an injected grout. [It also has been suggested that dry bentonite could be blown into this open space, with hydration occurring naturally (Jessberger, 1993a).] The grout must be quick-setting, so that it will support the weight of the overburden without undue deformations when the ''sword'' moves on. It also should have a low permeability in case the polyethylene sheet develops a defect. The strips of polyethylene are spliced by a robotlike machine that passes back and forth between the tubes and seals the seams by melt welding (see item A in Fig. 4.8). As shown in Fig. 4.10, the tubes provide access for the construction of the liner system, including the watertight connection of the liner to the tubes. Moreover, all steps of the installation can be observed and controlled from the tubes. The tubes also provide access for future inspection and monitoring and any repairs

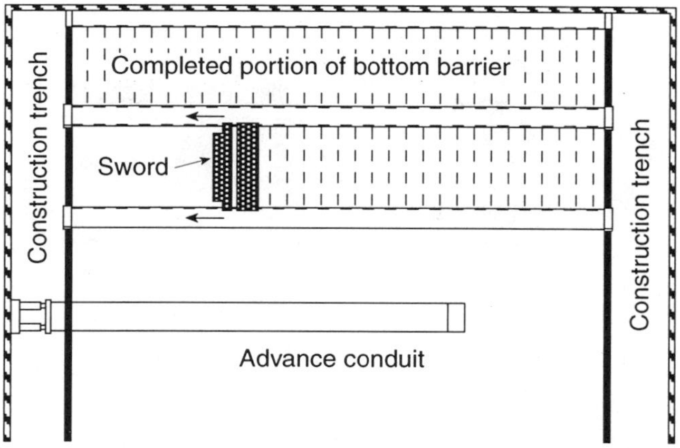

Figure 4.7a Step-by-step production of the bottom barrier (Jessberger, 1993b).

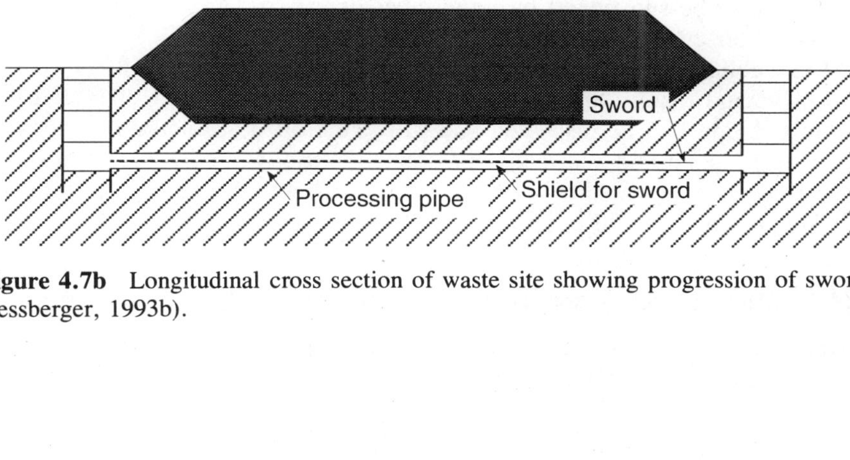

Figure 4.7b Longitudinal cross section of waste site showing progression of sword (Jessberger, 1993b).

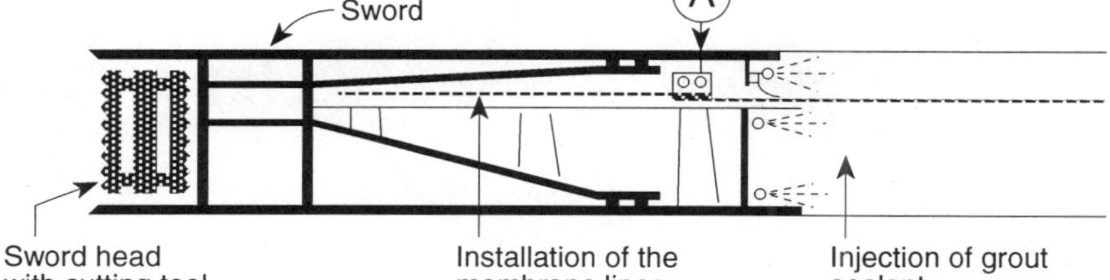

Figure 4.8 Longitudinal cross section of the sword (Jessberger, 1993b).

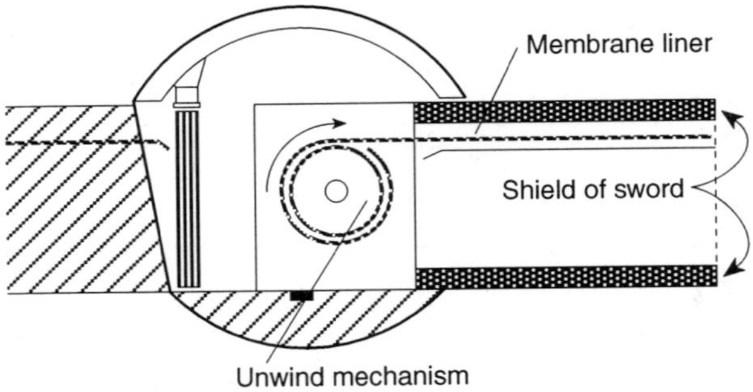

Figure 4.9 Membrane liner being pulled from conduit to conduit (Jessberger, 1993b).

that may be necessary. This would help to assure reliable long-term performance.

The sides of the containment system can be treated as in Fig. 4.11, where the wall (designated by B) is a vertical cutoff wall with a synthetic liner, C, spliced to the bottom barrier liner, D. If the side shafts have to be excavated below the water table in a highly permeable soil formation, then the open trenches may be substituted by large tubes or caissons from which the tubes that go under the waste are pushed.

The Zublin system is an untried high-technology system. Its use would be expensive and require highly skilled labor. On the other hand, the quality of the installation can be inspected and fully controlled. In addition, sampling tubes could be pushed below the constructed barrier from the large and accessible steel tubes to monitor the performance of the barrier system. If defects are found, repairs could be made from the large tubes.

4.4.2 Microtunneling and Injection Grouting

As in the Zublin system, the use of microtunneling involves pushing steel tubes into the soil from excavated trenches on the side of the waste site. These tubes

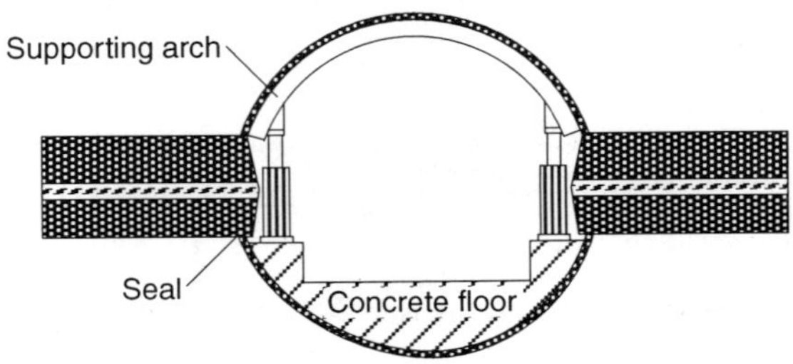

Figure 4.10 Cross section of an accessible construction conduit (conduit diameter = 2.8 m) (Jessberger, 1993b).

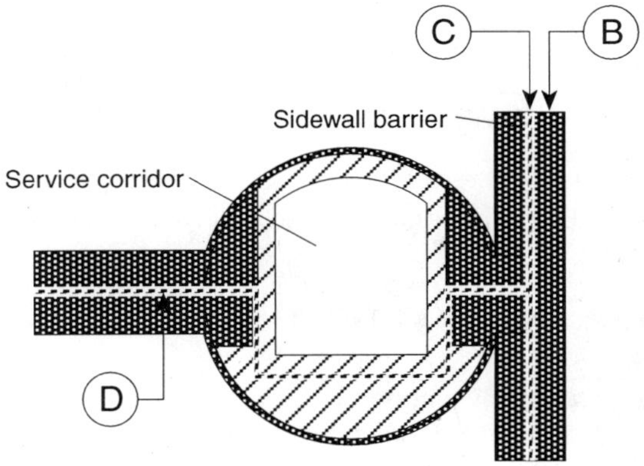

Figure 4.11 Steel conduit for servicing/control and connecting bottom barrier to sidewall barrier (Jessberger, 1993b).

are accessible to workers and are spaced depending on the groutability of the soil. Once the tubes are in place, injection rods are pushed from their sides toward adjacent tubes and are used to grout a bottom barrier between the tubes as shown in Fig. 4.12 (Meseck and Glaser, 1986). The steel tubes can be pushed as far as 450 ft. The alignment in this case is not as critical as in the Zublin system.

4.4.3 Rectangular Microtubes Pushed Side by Side

In this system rectangular steel tubes are pushed horizontally from a clean open trench on the side of the waste. The tubes are jacked side by side into the soil below the waste to provide a watertight bottom barrier as shown in Fig. 4.13 (Meseck and Glaser, 1986). The tubes are cleaned out as they are pushed and can either have an interlocking connection (like sheet piles), or the space between them can be grouted by permeation grouting. The tubes should be large enough for access by workers, if grouting is used.

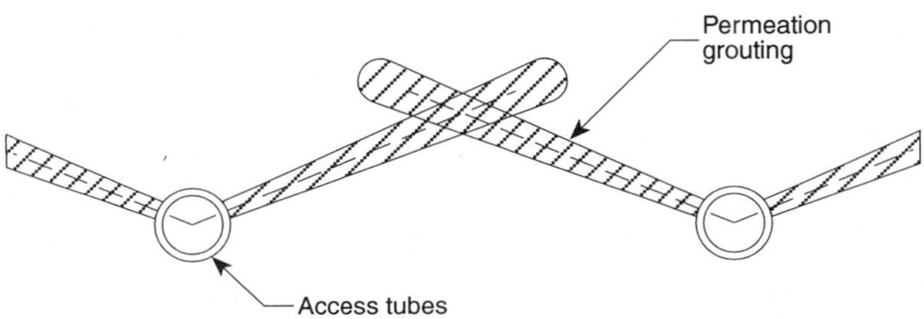

Figure 4.12 Bottom barrier constructed by permeation grouting from parallel access tubes (microtunnels).

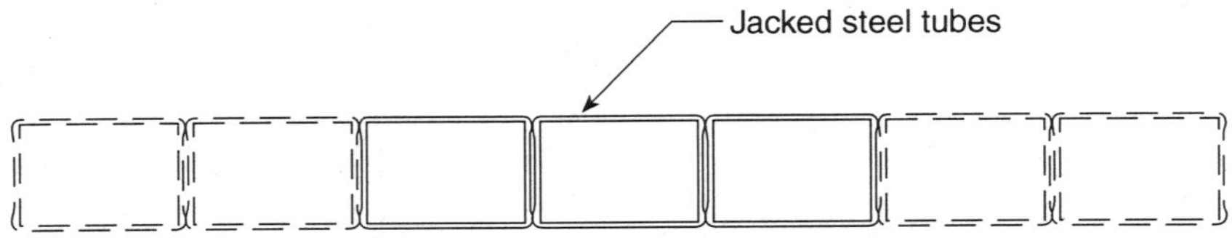

Figure 4.13 Bottom barrier constructed of rectangular steel tubes pushed side by side.

4.5 ECONOMIC AND PRACTICAL CONSIDERATIONS

In general, if an acceptable natural low-permeability clay layer or rock formation exists at a moderate depth beneath the waste, then this clay or rock should be used to create the bottom seal. Cutoff walls can be keyed into this natural clay or rock and construction of a bottom barrier becomes unnecessary.

The following considerations give guidance on selection of the most economical containment system. An idealized waste site of cylindrical shape has been assumed, with a radius *r* and a depth *D*, as shown in Fig. 4.14. The waste is assumed to be located in a highly pervious cohesionless soil. Also assumed is the depth *H* to a low-permeability bottom clay layer.

Two extreme containment situations are considered. In the first, a vertical barrier is used with a bottom barrier. The sidewall barrier is extended to a depth *D*. In the second case, the sidewall barrier extends all the way to the bottom clay and keys into it, so its depth is *H*. The cost of both of these cases has been computed. In doing this, it is assumed that the vertical barrier (side curtain) has a unit cost of \$(SC) per square foot, and the bottom barrier (seal) has a unit cost of \$(BS) per square foot.

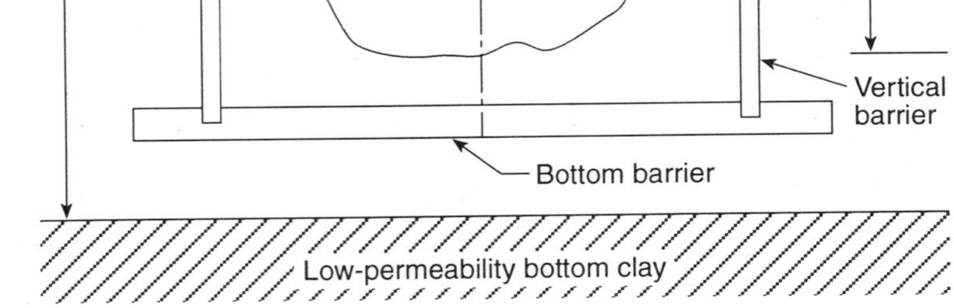

Figure 4.14 Cross section of a waste site with both vertical and bottom barriers (also shown is a low-permeability bottom clay).

TABLE 4.1 Variation of Critical Depth of Waste with Site Radius for a Clay Bottom Depth of 100 ft

For H = 100 ft:	
r (ft)	D_{cr} (ft)
50	75
100	50
150	25
200	0

Cost of case No. 1: $2(\pi)rD \times \$(SC) + (\pi)r^2 \times \$(BS)$

Cost of case No. 2: $2(\pi)rH(\text{clay}) \times \$(SC)$

It is important to calculate the critical depth of waste D_{cr}, where the cost of the two containment cases is the same. Once this is known, the less expensive of the two alternatives can be chosen. Namely, for a given depth to the clay bottom H, if the D of the waste is less than D_{cr}, then the combined use of a sidewall and bottom barrier would yield the cheaper alternative.

To find D_{cr}, the depth of the waste D will be set equal to D_{cr}, and the cost of the two alternatives will be calculated and set equal, thus; $D_{cr} = H - (r/2)(\$BS)/(\$SC)$.

As an example, if H = 100 ft, r = 150 ft, and \$BS = \$SC = \$50.00, then D_{cr} = 25 ft. So for a waste depth of 25 ft or less the combined use of a sidewall and bottom barrier is the more economical alternative.

Tables 4.1 and 4.2 show how D_{cr} varies with the radius of the waste site for assumed H values of 100 and 150 ft, respectively. For both the sidewall and the bottom barrier, a unit price of \$50.00 was assumed.

If in the above equation D_{cr} is set to its minimum (i.e., zero), the solution of the equation will give, as a function of H, the maximum radius of a waste site for which a combined sidewall and bottom barrier should be considered. Above that maximum radius the containment should be built without a bottom

TABLE 4.2 Variation of Critical Depth of Waste with Site Radius for a Clay Bottom Depth of 150 ft

For H = 150 ft:	
r (ft)	D_{cr} (ft)
50	125
100	100
150	75
200	50
250	25
300	0

TABLE 4.3 Variation of Maximum Site Dimensions and Area with Depth of Clay for Installation of Bottom Barriers

H (ft)	*r* (max) (ft)	Diameter (max) (ft)	Area (max) (sq ft)	Area (max) (acres)
50	20	40	1,257	0.03
100	40	80	5,027	0.12
150	60	120	11,310	0.26
200	80	160	20,106	0.46

barrier; that is, the sidewall barrier should be extended to the bottom clay or rock.

The following tables show how the maximum waste site radii and areas vary with the depth of the clay H, and the unit cost of the containment elements. For Table 4.3 a simple slurry wall with \$(SC) = \$10.00 and a jet grouted bottom seal with \$(BS) = \$50.00 was assumed.

Table 4.3 shows that with a relatively inexpensive vertical barrier only small waste sites would warrant the use of a bottom barrier. In other words, in many cases it would be more cost effective to extend the sidewall barrier all the way to the natural bottom clay or rock.

For Table 4.4 a more expensive slurry diaphragm wall and a jet grouted bottom barrier were assumed, both with a unit price of \$50.00/ft^2. Table 4.4 shows that with a more sophisticated (and expensive) vertical barrier, the construction of a bottom barrier becomes an economically viable and attractive alternative.

4.6 CONSTRUCTION QUALITY CONTROL

The concept of constructing a low-permeability bottom barrier within high-permeability soil or rock formations underlying waste deposits is relatively new compared to the construction of vertical sidewall barriers and surface covers. In fact, Walker (1993) contends that no such projects have been con-

TABLE 4.4 Variation of Maximum Site Dimensions and Area with Depth of Clay for Installation of Bottom Barriers

H (ft)	*r* (max) (ft)	Diameter (max) (ft)	Area (max) (sq ft)	Area (max) (acres)
50	20	200	31,400	0.7
100	200	400	125,600	2.9
150	300	600	282,740	6.5
200	400	800	502,650	11.5

structed in the United States to date, although such projects have been undertaken elsewhere in the world. Geological and hydrogeological characterization of a proposed site is the first step in construction quality control. A comprehensive and accurate site characterization will enhance the likelihood of constructing a successful bottom barrier beneath a waste site.

Visual inspection and laboratory testing, which are traditional elements of construction quality control at the surface, can be applied to the construction of surface covers (or caps) and, to some extent, to the construction of sidewall barriers. However, visual inspection and laboratory testing have limited applicability to subsurface construction of a bottom barrier using drilling technologies and permeation or jet grouting technologies. In this latter case, nondestructive testing and remote sensing and detection instrumentation must supplant the laboratory testing and visual observations common to construction at the surface. Reference has already been made in this chapter to remote measurement of the cavity created during jet grouting utilizing sonar instrumentation and to tracking the position of the drill head during directional drilling utilizing radio wave transmission and reception instrumentation. It is quite likely that successful quality control for bottom barrier construction will depend on continued development of nondestructive remotely-operated sensing, detection, and measuring technologies. An exception to this might be construction of a bottom barrier utilizing the Zublin technique in which direct visual observation and testing can be more easily carried out.

During subsurface grouting, using either permeation grouting or jet grouting, careful measurement can be made of the hydraulic pressure applied to the grout and to the volume of grout delivered. Unfortunately, they are only indirect measurements of the degree to which a continuous and low-permeability bottom barrier is being constructed; although, this type of information permits inferences to be made and can increase confidence that quality is being achieved. The watertightness of a grout barrier can be checked by field tests utilizing piezometers on both sides of the constructed barrier.

Recently, proposals have been advanced calling for the application of technologies, previously developed for other purposes, to construction quality control for bottom barriers (Witherspoon, 1993; Rector and Morrison, 1993). Witherspoon proposes that "tiltmeters," previously developed to measure small changes in the bulge of the earth, may be able to monitor the subsurface injection of grouting fluids during placement of a bottom barrier. The tiltmeter is capable of measuring changes in the tilt of the surface in nanoradians (10^{-9} radians). Davis (1983) has shown that the shape of the surface deformation has a very distinct signature that depends on the orientation of the subsurface deformations and volume changes.

Rector and Morrison (1993) propose using multiple cross-hole seismic and electrical methods to characterize grout injection before, during, and after placement. As described, both methods would permit geophysical verification of the continuity of an arbitrary size bottom grout barrier without access to the surface of the contaminated area. It is envisioned that complete electrical and seismic surveys would be performed before and after grout placement. The first survey would be used as the baseline against which the results of the second survey would be compared.

Whatever techniques and technologies are employed to measure the quality of the constructed bottom barrier, the detection of defects in the constructed barrier, provision for capture and recovery of leakage from the contained waste site, and repair of the barrier are issues that need to be addressed. These and other issues related to construction quality control related to bottom barriers are given further consideration in Chapter 6.

5 Surface Control Barriers (Caps)

5.1 INTRODUCTION

Engineered caps are widely used as part of the overall strategy for containment of buried wastes at waste disposal sites. Because of the variety of waste characteristics, available materials to construct a cap, and site conditions, caps must be tailored to the specific requirements of each particular project. The components of a final cap, and the best materials to use for each component, cannot be generalized for all sites. Each site must be evaluated on a case-by-case basis.

Depending on site-specific circumstances, caps may be called upon to perform one or more of the following functions:

- Raise the ground surface and provide appropriate slopes to promote runoff and controlled drainage of surface water.
- Separate buried waste from plants, animals, and humans.
- Minimize infiltration of water into the waste.
- Control the release of gas out of the waste.

The cap must perform the intended functions for an extended period of time. The required "design life" of a cap depends primarily on the nature of the waste, site conditions, and the likelihood of extended maintenance of the cap. A minimum design life may be specified by regulations.

The design of a final cap is complicated by factors such as:

- Temperature extremes, including possibility of freeze–thaw conditions.
- Precipitation extremes, which can lead to cyclic wetting and drying that can cause development of shrinkage cracks in some soils.
- Penetration of plant roots, burrowing animals, worms, and insects into soil.
- Inadvertent human excavations into the cap.
- Subsidence of the waste, which can change drainage contours and cause cracking in some soils.
- Down-slope slippage or creep, which can result in landslides or cracking.
- Vehicular movements on the cap, which may damage the cap.

- Deformations caused by earthquakes.
- Wind or water erosion of materials.

Design of an infiltration barrier is particularly challenging because the barrier may be subjected to wet–dry cycles and large differential settlement, and yet the barrier layer in a final cap is intended to maintain a very low permeability to water and gas for long periods of time.

In this chapter, the components of a final cap are identified, the potential materials that could be used are listed, advantages and disadvantages of various materials are given, the major problems and technical issues are described, and an overall assessment of caps is provided.

5.2 COMPONENTS OF A FINAL CAP

The possible components of a final cap are sketched in Fig. 5.1. From top to bottom, the five components consist of a surface or erosion layer, a protection layer, a drainage layer, a barrier layer, and a foundation and/or gas collection layer. Some layers may contain several materials, for example, a barrier layer may be composed of a geomembrane and a low-permeability compacted clay liner that together form a composite liner.

All five components are not needed for all caps. Many final caps may contain just two or three of the five possible components shown in Fig. 5.1. The only layer needed for all caps is the surface layer. Figure 5.1 also lists some of the materials typically used in final caps.

5.2.1 Surface Layer

The function of the surface layer is to separate the underlying components of the cap from the ground surface and to minimize temperature and precipitation extremes in underlying layers. The surface layer itself must resist wind and water erosion and, in general, be survivable. "Survivability" depends greatly on the amount of maintenance that can be guaranteed. Often, caps are designed on the assumption that minimal maintenance will be performed for a few years and after that, no maintenance is assumed.

Four materials have been used for the surface layer in a final cap: (1) topsoil; (2) geosynthetic erosion control material overlying topsoil; (3) cobbles; and (4) paving material. The relative advantages and disadvantages of each material, along with an overall assessment, is provided in Figs. 5.2–5.5 for each of these materials.

Of the four alternative materials for the surface layer, topsoil (Fig. 5.2) has been by far the most commonly used material. Topsoil is usually available from the site and locally at nominal cost. Suitable topsoil will promote growth of vegetation. Roots of plants reinforce the topsoil, reduce the rate of soil

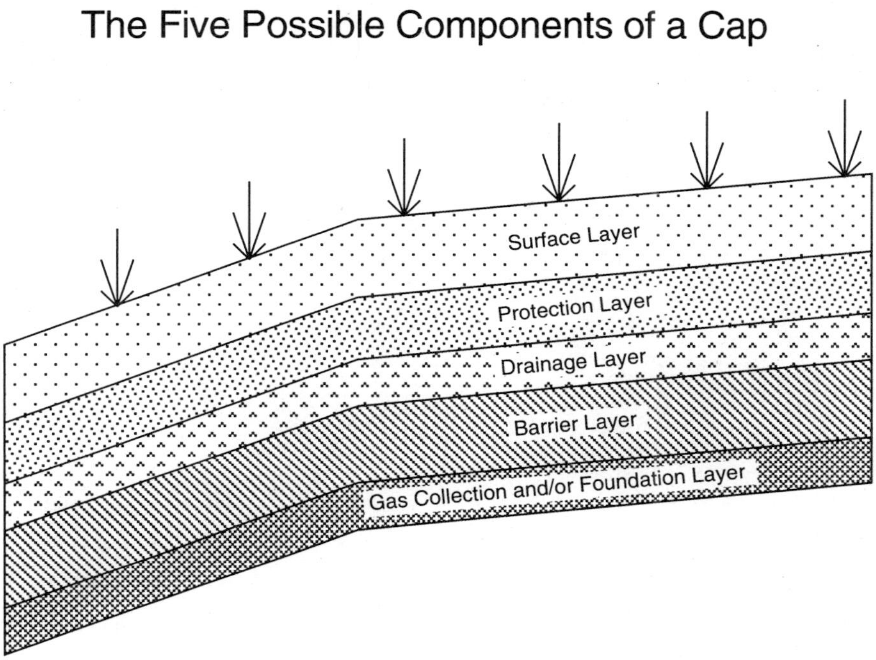

Layer	Description of Layer	Typical Materials	Typical Thickness (ft)
1	Surface Layer	Topsoil; Geosynthetic Erosion Control Layer; Cobbles; Paving Material	0.5
2	Protection Layer	Soil; Recycled or Reused Waste Material; Cobbles	1-3
3	Drainage Layer	Sand or Gravel; Geonet or Geocomposite	1
4	Barrier Layer	Compacted Clay; Geomembrane; Geosynthetic Clay Liner; Waste Material; Asphalt	1-3
5	Gas Collection Layer and/or Foundation Layer	Sand or Gravel; Soil; Geonet or Geotextile; Recycled or Reused Waste Material	1

Figure 5.1 The five basic components of a final cap.

Layer 1 -- Surface Layer

Alternate A: Topsoil

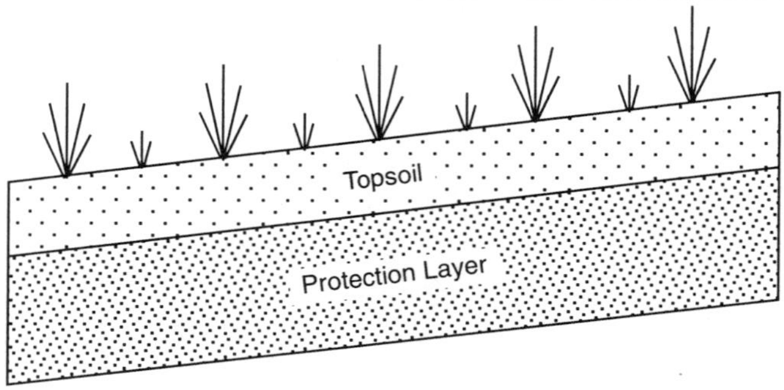

Advantages

1. Construction materials are usually available locally.
2. Topsoil stores water that has infiltrated; plants return much of this water to the atmosphere via evapotranspiration.
3. Plants not only return water to the atmosphere but also reduce soil erosion and stabilize the topsoil.
4. Plant roots tend to be confined mainly to the nutrient-rich topsoil (deep penetration of roots is undesirable).

Disadvantages

1. Requires adequate precipitation to support growth of vegetation on topsoil (design unsuitable for some arid sites).
2. Irrigation could be required during periods of drought to maintain vegetative growth.
3. Establishing growth of vegetation may be difficult if weather conditions are unfavorable immediately after seeding.
4. Excessive wind or water erosion may occur prior to establishment of vegetative cover or if vegetative cover becomes thin.
5. Erosion protection difficult for steep slopes.
6. Little resistance to intrusion by burrowing animals.

Assessment

This is the most common surface material used in caps and is the material required by many regulatory agencies. The main problems experienced with topsoil are maintaining a vegetative growth during drought periods and formation of erosion gullies during excessive precipitation periods or on steep slopes. This material can be very effective at sites that receive adequate rainfall and have moderate slopes, but topsoil generally requires maintenance in early years (before vegetative growth is fully established) to repair erosion gullies.

Figure 5.2 Topsoil used for surface layer.

Layer 1 -- Surface Layer

Alternate B: Geosynthetic Erosion Control Material Over Topsoil

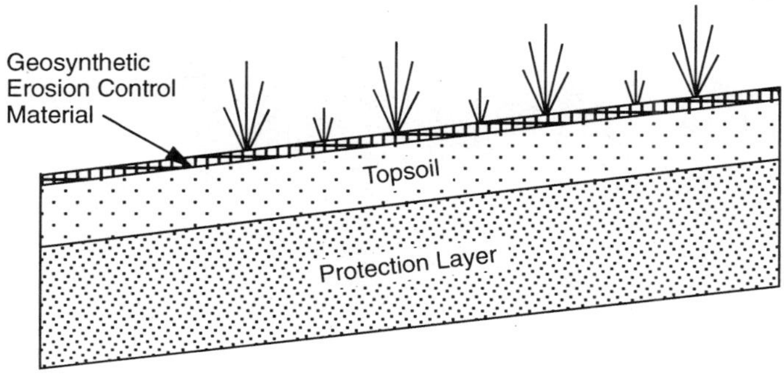

Advantages

1. Topsoil stores water that has infiltrated; plants return much of this water to the atmosphere via evapotranspiration.
2. Plant roots plus geosynthetic erosion control material minimize soil erosion; geosynthetic erosion control material minimizes erosion in early years while plant growth is established -- soil containing roots provides long-term protection from erosion.
3. Plant roots tend to be confined to nutrient-rich topsoil (deep penetration of roots is undesirable).
4. Geosynthetic erosion control material offers some resistance to burrowing animals.
5. Geosynthetic erosion control material can be anchored or reinforced to add stability to steep side slopes.
7. The geosynthetic erosion control material can be preseeded to aid with establishment of vegetative growth.

Disadvantages

1. Requires adequate precipitation to support growth of vegetation on topsoil (design unsuitable for some arid sites).
2. Irrigation could be required during periods of drought to maintain vegetative cover.
3. Establishing growth of vegetation may be difficult if weather conditions are unfavorable immediately after seeding.

Assessment

This design is not nearly as common as the same design without the geosynthetic erosion control layer. However, many erosion problems have been observed on covers; the erosion control layer is particularly useful for sites with steep slopes.

Figure 5.3 Geosynthetic erosion control material used for surface layer.

Layer 1 -- Surface Layer

Alternate C: Cobbles Used for Surface Layer

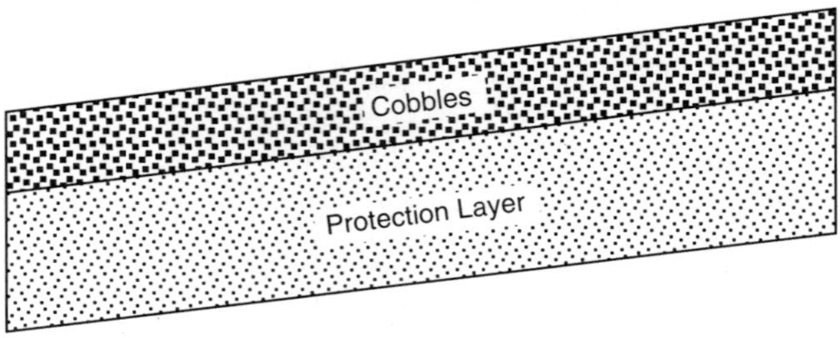

Advantages

1. No need to establish or maintain vegetative growth (particularly advantageous for arid sites or sites subject to periodic droughts).
2. Cobbles provide protection from wind or water erosion; effectiveness of cobbles not dependent upon maintenance of vegetative growth.
3. Cobbles may minimize penetration from burrowing animals if rocks are of adequate size.
4. Cobbles minimize growth of plants with deep roots that could penetrate underlying barrier layers.

Disadvantages

1. Cobbles must be adequately large to resist washout; materials may not be readily available locally.
2. Minimal evaporation of water from underlying soil (cobbles allow water to percolate downward but insulate soil from surface evaporation).
3. Wind-blown materials may fill voids between cobbles; deeply-rooted plant growth may eventually become established and require maintenance.

Assessment

Cobbles are usually considered only for arid sites where it is difficult to maintain a growth of vegetation for erosion protection. A serious shortcoming with this design, however, is that the cobbles allow rainfall to infiltrate but do little to promote evapotranspiration of water from the subsoils back to the atmosphere.

Figure 5.4 Cobbles used for surface layer.

Layer 1 -- Surface Layer

Alternate D: Paving Material

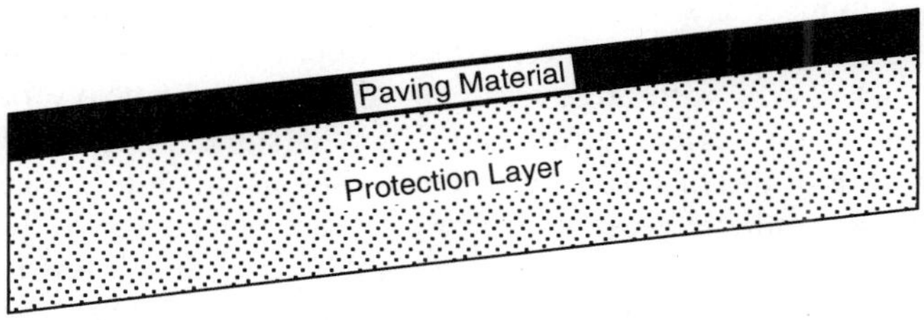

Advantages	Disadvantages
1. Construction materials are usually available locally. 2. May be able to use cover for parking of vehicles. 3. Surface layer also serves at least partly as hydraulic barrier. 4. Extremely high resistance to erosion. 5. Resistance to penetration by burrowing animals is provided.	1. Paving will require maintenance. 2. No evaporation or evapotranspiration of water occurs from protection layer. 3. Subsidence will cause cracking in the paving material. 4. Pavement will eventually degrade.

Assessment

This design is usually considered only for projects where the subsurface materials provide stable foundation support, the paving material can be maintained, and the surface layer need only function for a limited period before it is maintained or replaced (e.g., with a pavement overlay).

Figure 5.5 Paving material used for surface layer.

erosion, decrease runoff velocity, and remove water from the soil via evapotranspiration. Vegetation also creates a leaf canopy above the soil to reduce rainfall impact and decrease wind velocity.

The main problem experienced with topsoil used for a surface layer has been excessive erosion, especially on newly seeded caps or steeply sloping caps. For example, Swope (1975) studied 24 landfill caps in Pennsylvania and found that 33% had slight erosion, 40% moderate erosion, and more than 20% had severe erosion. Johnson and Urie (1985) report that erosion is made more severe from the installation of hydraulic barriers (which, without a drainage layer, can produce soaked soil with large amounts of runoff after heavy rains) and vehicles driving on the surface of the cap. Many designers underestimate the problem of erosion of the surface layer prior to establishment of a mature vegetative cap. A geosynthetic erosion control material placed above the topsoil (Fig. 5.3) can be very beneficial in limiting problems with soil erosion in the period following installation and can even help to establish vegetative growth with seeds (even on a time-release basis) embedded in the layer. The USEPA (1983) and Gilman et al. (1985) provide guidance for planting vegetation on final caps.

A problem occasionally experienced with caps is lack of moisture in the topsoil, which makes maintenance of vegetation very difficult. The problem can be acute when very sandy topsoil is used and a drainage layer is located within the cap. In some extreme cases, owner–operators have been forced to irrigate the final cap in order to maintain vegetative growth on the cap as required in the permit for the facility. The solution to the problem is to employ topsoil with sufficient capacity to retain water.

Additionally, the problem of stability of the topsoil must be addressed. A number of failures of caps were reported by Boschuk (1991). Most of these failures were attributed to fundamental errors made in assessing the stability of the cap, or in several instances, lack of an analysis of the stability of the cap. Geogrid or geotextile reinforcement layers can be incorporated in the cap to increase factors of safety against slope instability to acceptable levels. In particular, critical interfaces along which slippage would most likely occur will be described later.

In areas where the amount of precipitation is inadequate to support growth of a vegetative layer, an alternative is to armor the surface with cobbles (Fig. 5.4). However, cobbles serve as a one-way window for water, that is, precipitation can infiltrate downward through the cobbles, but due to lack of plant roots, little water is returned to the surface via evaporation. In studies at Hanford, Washington, researchers found that native soil materials in field lysimeters allowed no percolation through a native soil, but when the ground surface was capped with gravel, approximately one-half of the rainfall (average of 6 in. annually) percolated through the prototype caps contained in the lysimeters. Thus, cobbles for a surface layer may be undesirable in terms of limiting infiltration of water, even in desert climates.

Asphalt concrete surfaces (Fig. 5.5) have been used on occasion for the surface layer (Repa, 1987). At the Oak Ridge Y-12 facility, an asphalt concrete cover layer was placed over a closed pond—drainage and barrier layers were located beneath the pavement. The pavement was used as a parking lot. The main problems with asphalt concrete are that any significant differential settlement will cause cracking, and the pavement will have a limited lifetime. Inyang and Myers (1993) offered guidance on other types of surface functional structures, particularly for RCRA sites.

5.2.2 Protection Layer

A protection layer may serve several functions:

- To store water that has infiltrated into the cap until the water is later removed by evapotranspiration.
- To physically separate the waste from burrowing animals or plant roots.
- To minimize the possibility of human intrusion into the waste.
- To protect underlying layers in the cap from excessive wetting–drying, which could cause cracking of some materials.
- To protect underlying layers in the cap from freezing which could also cause cracking in some materials.

Figures 5.6–5.8 illustrate some of the alternative materials that are used for the protection layer. Local soil (Fig. 5.6) is by far the most commonly used material. Soil can be very effective in storing water for later evapotranspiration and provides protection against freezing of underlying materials if the thickness of the protection layer is sufficiently large. In many cases, the surface and protection layers are combined to form a single "cap soil."

Waste materials could serve as the material for the protection layer (Fig. 5.7). Materials such as fly ash, bottom ash, and paper mill sludge, could be used for this layer.

Cobbles (Fig. 5.8) have been used occasionally as a barrier to plant roots and burrowing animals at some types of waste disposal facilities. The primary use for the "biotic barrier" of cobbles has been for protection of radioactive waste, which is known to have been ingested by burrowing animals at some sites (Cline et al., 1982).

5.2.3 Drainage Layer

A drainage layer is sometimes placed below the protective layer and above the barrier layer (Fig. 5.1). There are three reasons why a drainage layer might be desirable:

1. To reduce the head of water on the underlying barrier layer, which minimizes percolation of water through the cap.

Layer 2 -- Protection Layer

Alternate A: Soil for Protection Layer

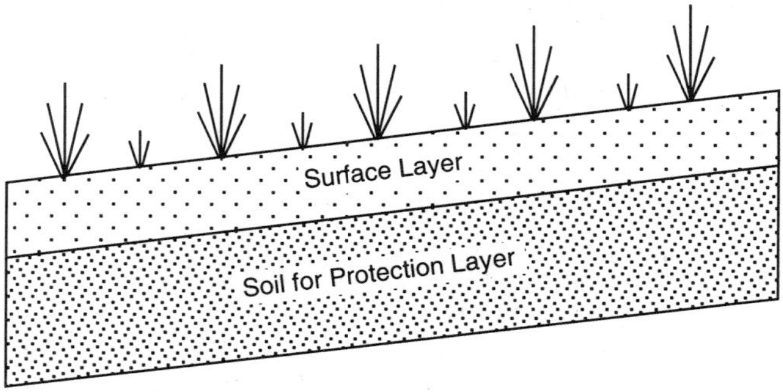

Advantages	Disadvantages
1. Construction materials are usually available locally. 2. Soil stores water that has infiltrated; plants return much of this water to the atmosphere via evapotranspiration.	1. Little resistance to intrusion by plant roots or burrowing animals.

Assessment

This is the most commonly used material for the protection layer used in caps. Except for cases where special design requirements exist for the protection layer, or alternative materials are readily available, soil will generally prove to be the most suitable material for the protection layer.

Figure 5.6 Soil used for protection layer.

Layer 2 -- Protection Layer

Alternate B: Recycled or Reused Waste Material for Protection Layer

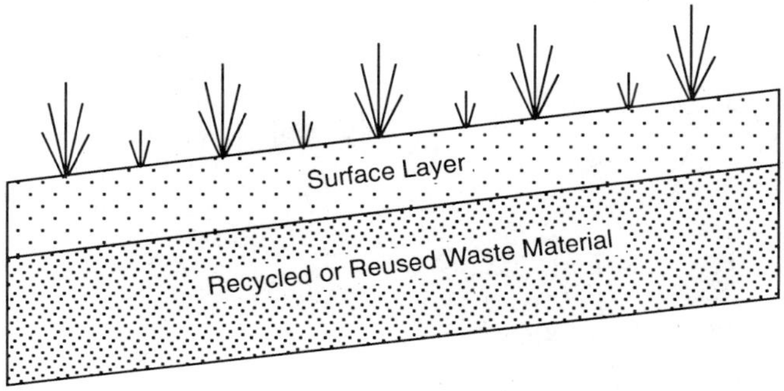

Advantages	Disadvantages
1. Offers productive use of waste materials. 2. Waste material may store water that has infiltrated; plants return much of this water to the atmosphere via evapotranspiration. 3. Some waste materials may offer resistance to penetration by plant roots and burrowing animals.	1. Suitable materials may not be available locally, leading to very large transportation costs 2. Not all recycled/reused waste materials are suitable for a final cover -- difficulty may be encountered in locating suitable materials.

Assessment

Waste materials have rarely been used in final caps, but the protection layer is one of the layers where recycled or reused materials show the greatest promise for serving as part of a final cap.

Figure 5.7 Recycled or reused waste material used for protection layer.

Layer 2 -- Protection Layer

Alternate C: Cobbles with Geotextile Filter

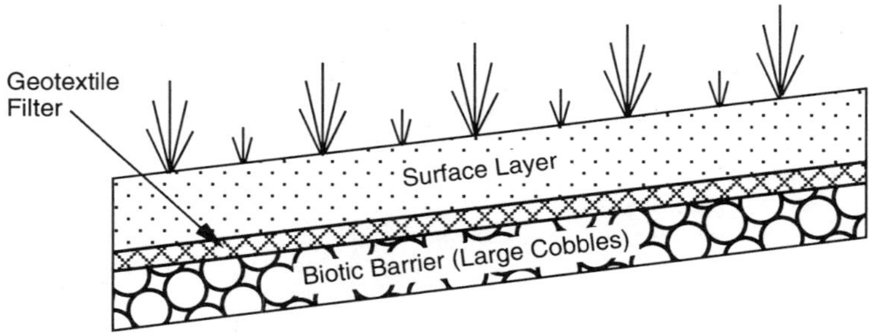

Advantages

1. Cobbles buried in the cover (biotic barrier) minimize penetration from burrowing animals and plant roots if rocks are of adequate size.

Disadvantages

1. Minimal water storage in cobbles (water tends to percolate downward rather than be stored for later evapotranspiration).
2. Cobbles may not be available locally.
3. Transportation costs for cobbles could be extremely large if cobbles are not available locally.

Assessment

Cobbles are usually considered only for arid sites requiring maximum security. A biotic barrier has rarely been used, except for radioactive waste buried at arid sites.

Figure 5.8 Cobbles used in protection layer as a barrier to burrowing animals.

2. To drain the overlying protection and surface layers, which increases the water-storage capacity of these layers and helps to minimize erosion of these layers by reducing the time during which the surface and protection layer materials remain saturated with water.
3. To reduce pore water pressures in the cap materials, which improves slope stability.

The materials used for drainage layers include sand, gravel with a filter, geonet with a geotextile filter, and geocomposite drainage material. These configurations are sketched in Figs. 5.9–5.12, and the advantages and disadvantages of each material are listed. There are other combinations of materials that are possible. For example, Fig. 5.10 shows a gravel drainage layer overlain by a geotextile filter, but the filter could be another material, for example, soil, and a geotextile cushion might be included at the base of the gravel layer to protect an underlying geomembrane in the barrier layer from puncture.

Each type of material can perform satisfactorily if properly designed and constructed. Therefore, selection of the type of drainage material is usually based upon economics and other considerations, such as ease and speed of construction. If the weight of the cover is an issue (e.g., if the cover tends to compress and consolidate underlying wastes or soils), geosynthetic materials enjoy an important advantage over thicker earthen materials.

Water must discharge freely from the drainage layer at the base of the cap (Fig. 5.13). If the outlet plugs or is of inadequate capacity, the toe of the slope may become saturated, develop excess pore water pressure, and potentially become unstable. Drainage pipes must have adequate capacity and cannot be allowed to plug. As mentioned earlier, the designer must be aware of the possibility that the drainage layer may drain overlying soils so well that irrigation is needed to maintain growth of vegetation. This problem is minimized by selecting surface layer materials and protection layer materials that have adequate water-holding capacity.

5.2.4 Barrier Layer

The barrier layer is often viewed as the most critical engineered component of the final cap. The barrier layer minimizes percolation of water through the cap by impeding infiltration through it and by promoting storage or drainage of water in the overlying layers. If the system is working properly, all or nearly all of the water that infiltrates into the cover system is removed by evapotranspiration or internal drainage. The barrier layer also restricts upward movement of any gases or volatile constituents that might be emitted by the waste.

Two materials have commonly been used for the barrier layer: compacted clay (bentonite) and geomembranes (Figs. 5.14 and 5.15). A relatively new material termed "geosynthetic clay liner," or GCL, is receiving strong consideration as an alternative to compacted clay. Section 5.3 discusses GCLs in

Layer 3 -- Drainage Layer

Alternate A: Sand Drainage Layer

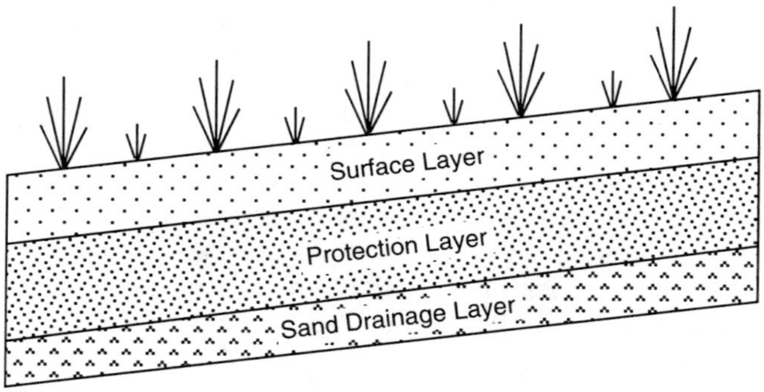

Advantages

1. Sand drainage layer helps to protect underlying layers from intrusion, puncture, and temperature extremes such as freeze/thaw .
2. Long history of use of sand drainage materials and sand in engineering applications.
3. May not need a filter if sand meets filter criteria to protect from intrusion by overlying protective layer.

Disadvantages

1. Suitable materials may not be readily available locally.
2. Fines in sand may tend to migrate downslope, which can lead to buildup of unreleased seepage pressures and slope instability.
3. May require a filter to protect from intrusion by overlying protection layer.
4. Sand may not have a suitably high hydraulic conductivity to serve the intended drainage function.

Assessment

This conventional material is technically sound, but its applicability can be limited by lack of suitable construction materials and a need at some sites to convey large flow quantities.

Figure 5.9 Sand used for drainage layer.

Layer 3 -- Drainage Layer

Alternate B: Gravel Drain with Geotextile Filter

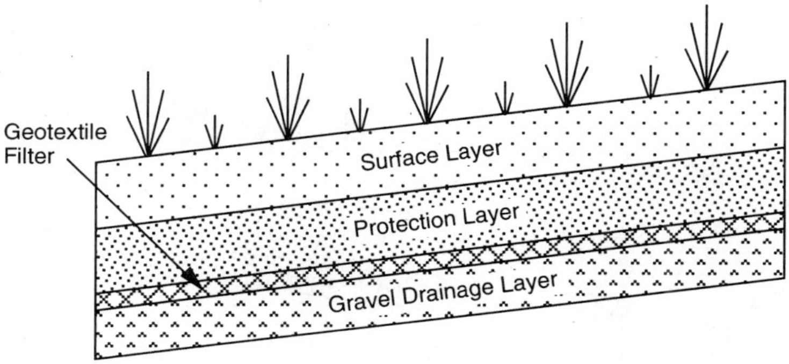

Advantages

1. Thick gravel drainage layer helps to protect underlying layers from temperature extremes such as freeze/thaw and from puncture.
2. Long history of use of gravel drains in engineering works.
3. Extensive history of use of geotextile filters to prevent clogging of granular drainage layers.
4. Geotextile filters are readily available and can be shipped to any location.

Disadvantages

1. Suitable gravel drainage materials may not be available locally.
2. Gravel drainage layer may tend to puncture underlying geomembrane or geosynthetic clay liner; a cushioning layer may be needed or restrictions on maximum particle size of gravel may be required.

Assessment

This conventional material is technically sound. Applicability may be limited by lack of suitable drainage materials in the local area.

Figure 5.10 Gravel with geotextile filter used for drainage layer.

Layer 3 -- Drainage Layer

Alternate C: Geonet Drain with Geotextile Filter

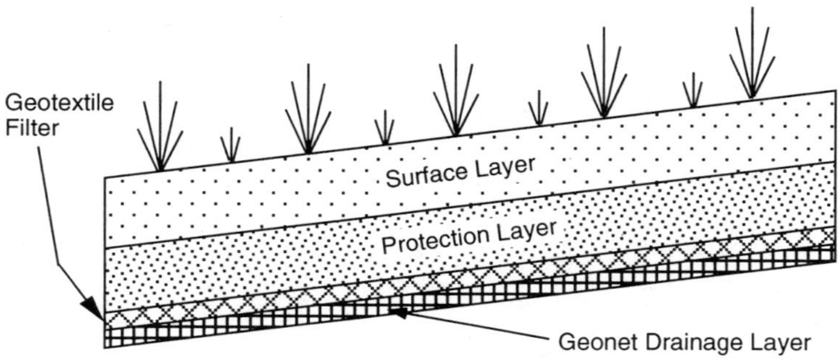

Advantages

1. Geotextiles and geonets are readily available and can be shipped to any location.
2. Lightweight installation epuipment can be used; risk of damaging underlying liner during installation of drainage layer is minimal.
3. Weight of drainage material is minimal; this reduces settlement of underlying waste or soil.

Disadvantages

1. Thin drainage layer does not help very much to protect underlying materials from freeze/thaw effects.
2. Potential slippage along geotextile/ geonet interface; geotextiles bonded to geonets are preferred but long-term stability of bond must be assured. Geotextiles (bonded to geonet) can be anchored to improve stability on slopes.
3. Potential slippage along geonet/ geomembrane interface, if underlying liner contains a geomembrane on the upper surface.
4. Seaming of geonets on long slopes is not straightforward.

Assessment

This is a conventional design for a drainage layer that is placed directly on a geo-membrane liner. If suitable drainage materials are not available locally, this is often the best design alternative for a drainage layer. The design can also be used for a drainage layer placed directly on a compacted clay liner, but a sand or geotextile filter would have to separate the geonet from the compacted clay liner.

Figure 5.11 Geonet drain with geotextile filter used for drainage layer.

Layer 3 -- Drainage Layer

Alternate D: Geocomposite Drain

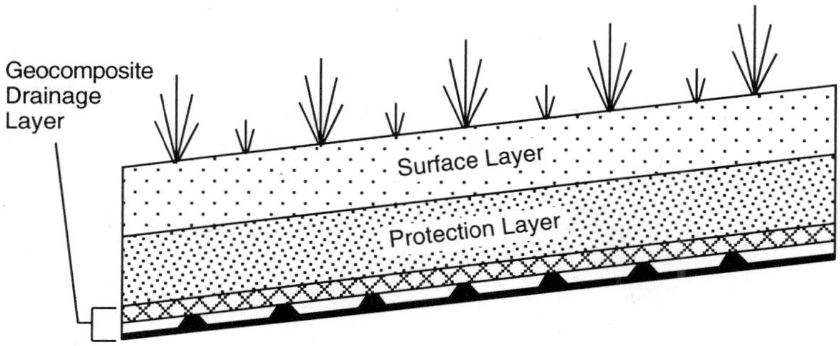

Advantages

1. Geocomposite drainage materials are readily available and can be shipped to any location.
2. Lightweight installation equipment can be used; risk of damaging underlying liner during installation of drainage layer is minimized.
3. Can be used on top of a compacted clay liner without need for an additional separation layer.
4. Connection of geocomposite cores is generally via a male/female interlock, which is very straightforward.
5. Light weight of geocomposite minimizes settlement of underlying waste or soil.

Disadvantages

1. Thin drainage layer does not help very much to protect underlying materials from freeze/thaw effects.
2. Potential slippage along geotextile/geocomposite interface. Bonded geotextiles to geocomposites are preferred, but long-term stability of bond must be assured.
3. Potential slippage along geocomposite/geomembrane interface, if underlying liner contains a geomembrane on the upper surface. Geotextile (bonded to geocomposite) can be anchored to assist stability.

Assessment

This design has not been used as frequently as a geonet drainage layer and geotextile filter. Technically, however, the geocomposite drainage layer can have much greater transmissivity, if required, than the geonet/geotextile combination.

Figure 5.12 Geocomposite drain used for drainage layer.

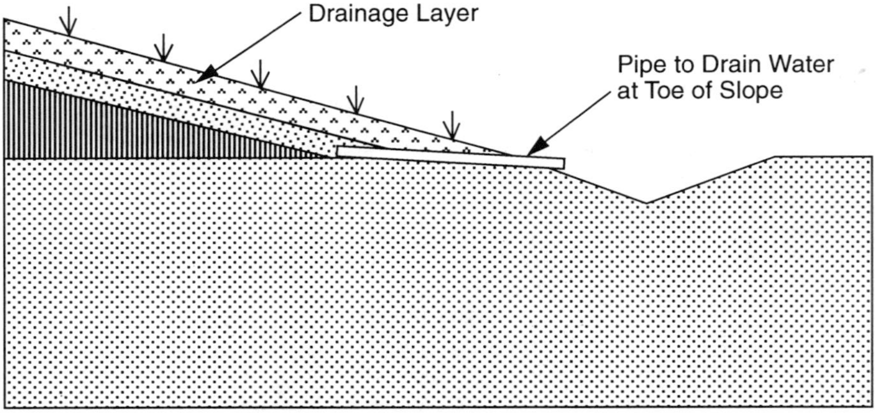

Figure 5.13 Pipe at toe of slope to drain water (pipe must not plug).

detail. A sketch of a GCL in a cap is shown in Fig. 5.16. In addition, composite geomembrane–clay liners have also been used, for example, a composite geomembrane–compacted clay liner (Fig. 5.17) or a composite geomembrane–GCL liner (Fig. 5.18).

Historically, a compacted clay liner (CCL) with a thickness of approximately 2 ft and a design hydraulic conductivity of no more than 1×10^{-7} cm/s has been the design standard for landfill caps. Much has been written about compacted clays (e.g., Daniel, 1987; Daniel, 1993a); this information is not repeated here. There are several potential problems that can make the long-term performance of a CCL questionable.

1. Clay liners are difficult to compact on highly compressible waste material; a suitable foundation layer is needed, and even then the compaction process can be difficult.
2. Compacted clay will tend to desiccate from above and/or below and crack unless adequately protected.
3. Compacted clay is vulnerable to damage from freezing and must be protected from freezing by a suitably thick layer of cap soil.
4. Differential settlement of underlying compressible waste will crack compacted clay if tensile strains become excessive.
5. Compacted clay liners are difficult to repair if they are damaged.

Some designers have assumed that a compacted clay liner can be protected from desiccation with a thin layer of cover soil. Experience illustrates the difficulty in protecting a clay from desiccation. Montgomery and Parsons (1989) describe field experiments in which three test plots were constructed at a site in Milwaukee, WI, and monitored for 3 years. A 4-ft thick layer of compacted

Layer 4 -- Barrier Layer

Alternate A: Single Compacted Clay Liner

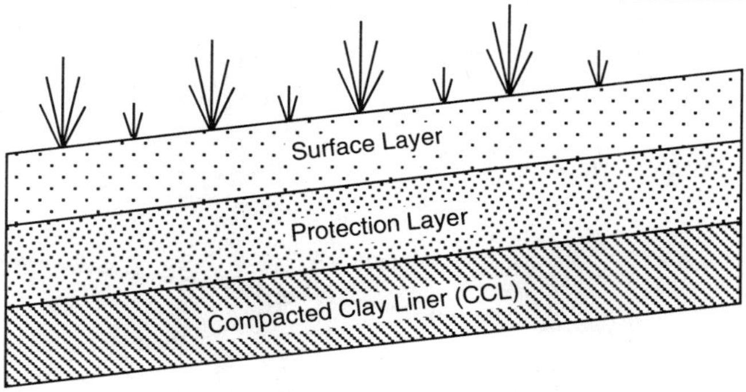

Advantages

1. Convenient construction if clay is available locally.
2. If clays are not locally available, local soils can be blended with imported, processed clay, e.g., bentonite.
3. Conventional method and material with long history of use.

Disadvantages

1. Clay may dry out from below and crack (dry waste or gas).
2. Clay may dry out from above and crack (arid climate, drought conditions, or thin cover soil).
3. Tension cracks may form as a result of differential settlement.
4. Freeze/thaw may damage liner (cold regions, thin soil cover).
5. Limited hydraulic effectiveness due to lack of composite action.
6. Clay may be difficult to compact above compressible waste.
7. Clay may not be available locally.
8. Liner is difficult to repair if cracked.
9. Stability concerns for steep slopes.

Assessment

The main problems with this liner are that it almost certainly will be damaged by wet/dry cycles and (depending on site conditions) may also be damaged by settlement and/or freeze/thaw. The only way to prevent wet/dry and freeze/thaw damage is to use an extremely (and often impractical) thick layer of cover soil. A single compacted clay liner is not recommended for most cover systems and should only be considered where there will be a very thick cover soil, no significant differential settlement, and no drying from below.

Figure 5.14 Single compacted clay liner used for barrier layer.

Layer 4 -- Barrier Layer

Alternate B: Single Geomembrane Liner

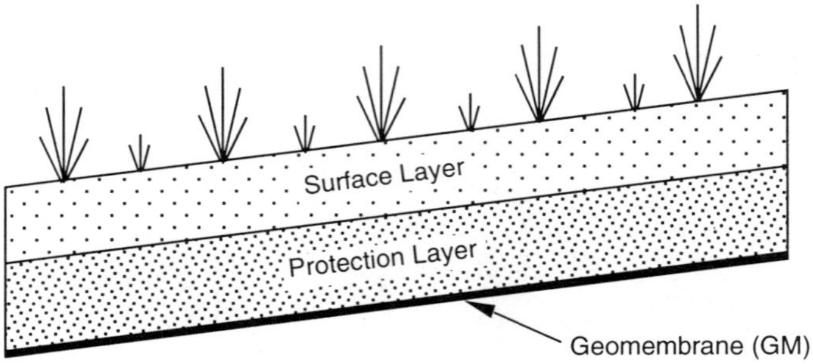

Advantages

1. Straightforward and rapid installation.
2. Materials are readily available and can be shipped to any location.
3. Liner not damaged by wet/dry cycles; does not require a thick cover soil to prevent desiccation of liner.
4. Geomembranes and their seams do no appear to be affected by freeze/ thaw cycling.
5. Some geomembranes can withstand very large differential settlement.
6. Easy to repair.
7. Gas collection penetrations are easy to construct (connection with a pipe boot).
8. Geomembrane can be anchored if stability is of concern.

Disadvantages

1. Potential slippage at interface between geomembrane and cover soil.
2. Tensile stresses may be mobilized in anchored geomembrane.
3. Vulnerable to puncture.
4. Limited hydraulic effectiveness due to lack of composite action.
5. Type and placement of cover soils is critical to minimize punctures.

Assessment

The main problem with this liner is that a geomembrane can be punctured during construction and later from accidental intrusion. This liner is only recommended when no other liner is practical and where some leakage through the liner can be tolerated.

Figure 5.15 Single geomembrane liner used for barrier layer.

Layer 4 -- Barrier Layer

Alternate C: Single Geosynthetic Clay Liner

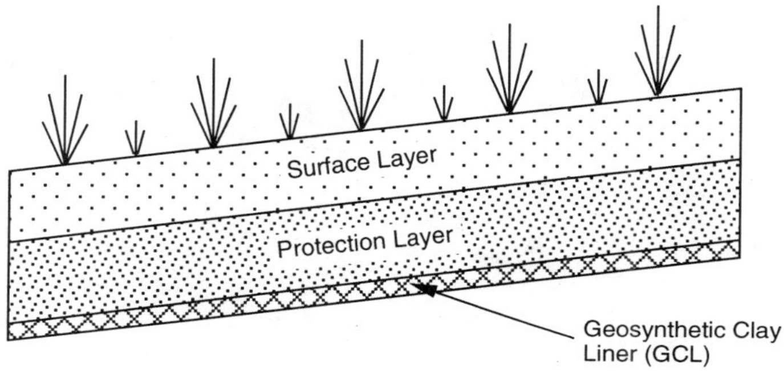

Advantages

1. Straightforward and rapid installation.
2. Materials are readily available and can be shipped to any location.
3. Geosynthetic clay liner possesses some self-healing capability from minor punctures, desiccation, and freeze/thaw (better in this regard than a compacted clay liner).
4. Gas collection penetrations are easy to construct.
5. Material is dry when placed and can be constructed in dry conditions without desiccation damage.
6. Geosynthetic clay liner can be easily repaired by patching if damaged.
7. Geosynthetic clay liner can be anchored if stability is of concern.
8. Geosynthetic clay liner can be reinforced if tensile stresses are of concern.

Disadvantages

1. Low shear strength of hydrated clay at mid-plane of some geosynthetic clay liners.
2. Potential slippage at interface between GCL and underlying filter.
3. Vulnerable to puncture.
4. Limited hydraulic effectiveness due to lack of composite action.
5. Placement of cover soil is critical to minimize punctures.
6. Dry GCL will not serve as a barrier to gas migration.

Assessment

Permeability of dry GCL to gas makes this liner alternative unacceptable for wastes that produce gas, unless one can be assured that the clay will be wetted soon after installation and will remain wet thereafter. The geosynthetic clay liner may be punctured during construction or after installation from accidental intrusion. This liner is only recommended for special cases where some leakage through the liner can be tolerated.

Figure 5.16 Single geosynthetic clay liner used for barrier layer.

Layer 4 -- Barrier Layer

Alternate D: Composite Geomembrane/ Compacted Clay Liner

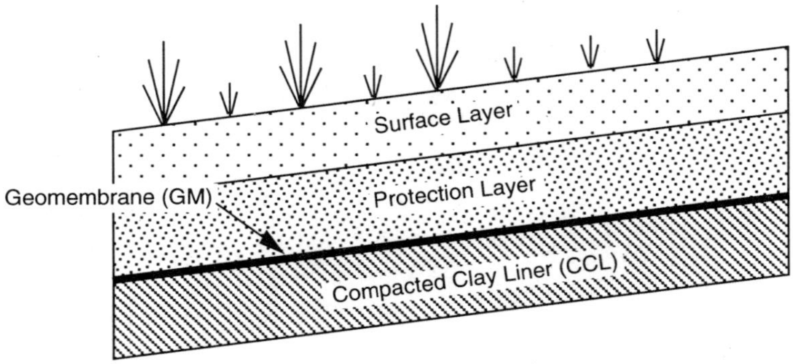

Advantages

1. Composite liner very effective as a hydraulic barrier.
2. Minimum liner required by some regulatory agencies.
3. Geomembrane protects clay liner from desiccation if cover soils dry; does not require a thick cover soil to prevent desiccation of clay.
4. Thick liner cannot be punctured.
5. Geomembrane component can tolerate large differential settlement.
6. Geomembrane can be anchored to provide stability.

Disadvantages

1. Potential slippage along both soil/ geomembrane interfaces.
2. Anchored geomembrane can cause tensile stresses to develop in the geomembrane.
3. Clay may dry from below and crack (dry waste or gas).
4. Tension cracks may form in clay as a result of differential settlement.
5. Freeze/thaw may damage clay liner (cold regions, thin soil cover).
6. Clay may be difficult to compact above compressible waste.
7. Clay may not be available locally.
8. Clay may not be easily repaired if damaged.

Assessment

A geomembrane/compacted clay liner is a common minimum requirement by various regulatory agencies for certain categories of waste. The primary concerns for this type of liner are with the long-term performance of the compacted clay component. Large differential settlement and drying from below can cause the compacted clay to crack and increase its permeability. However, if the geomembrane remains largely intact and the permeability of the compacted clay does not increase excessively, the composite liner would still be expected to be reasonably effective.

Figure 5.17 Composite geomembrane–compacted clay liner used for barrier layer.

Layer 4 -- Barrier Layer

Alternate E: Composite Geomembrane/ Geosynthetic Clay Liner

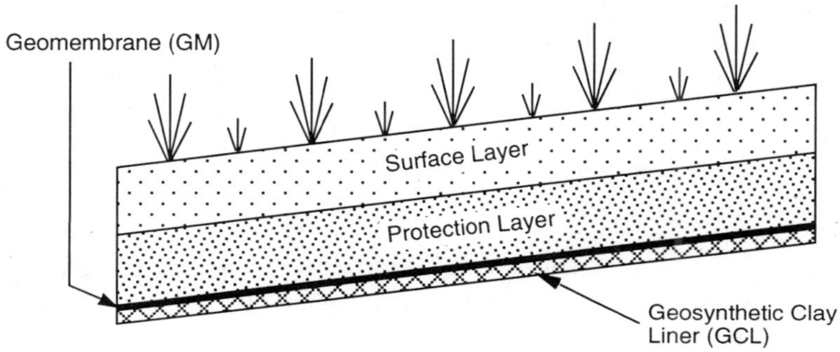

Advantages

1. Relatively straightforward and rapid construction.
2. Materials are readily available and can be shipped to any location.
3. GCLs possess some self-healing capability from minor punctures, desiccation, and freeze/thaw.
4. GCL is dry when placed and can be constructed in dry conditions without desiccation damage.
5. Liner can be repaired fairly easily.
6. Composite liner very effective as a hydraulic barrier.
7. Geomembrane component can tolerate large differential settlement.
8. GCL can be anchored if stability is a concern.
9. GCL can be reinforced if tensile stresses are of concern.

Disadvantages

1. Composite action questionable for most GCLs with upper geotextile.
2. Potential slippage along upper and lower surface of geomembrane and between some types of GCL and filter.
3. Low shear strength of hydrated clay at mid-plane of GCL for some products.
4. Vulnerable to puncture.
5. Placement of cover soil is critical to minimize punctures.
6. Differential settlement can cause clay deformation that could jeopardize low permeability of GCL in certain locations.

Assessment

This liner system offers redundancy and many practical benefits in terms of availability of materials and ease of construction. Questions can be raised about effectiveness of composite action between geomembrane and those GCLs with an upper geotextile -- this is an area of research at this time. Provided that issues concerning stability and puncture are resolved, this system probably offers the best overall benefit/cost ratio of any combination of barrier materials for cover systems.

Figure 5.18 Composite geomembrane–geosynthetic clay liner used for barrier layer.

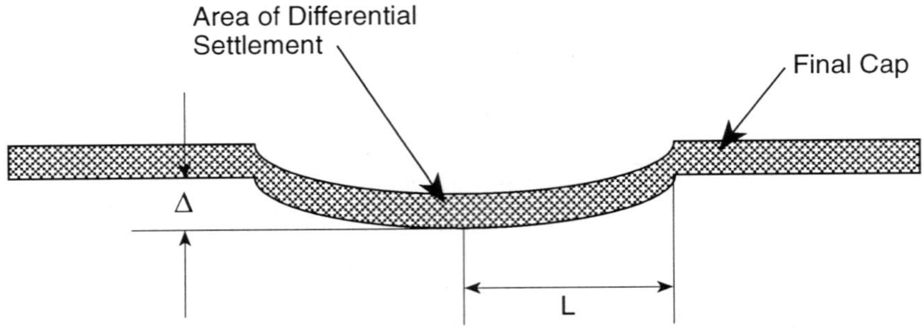

Figure 5.19 Definition of distortion, Δ/L.

clay was capped with either 6 or 18 in. of topsoil in separate test plots. After 3 years, excavations were made into the compacted clay. The condition of the clay liner in the two plots with 6 and 18 in. of topsoil was about the same. The upper 8–10 in. of clay was blocky, cracks up to 0.5 in. wide extended up to 38 in. into the clay, roots penetrated up to 10 in. into the clay in a continuous mat, and some roots extended up to 30 in. into the clay. Neither 6 or 18 in. of topsoil was enough to protect the clay adequately at this site.

Corser and Cranston (1991) describe test plots at a site near Kettleman City, CA, in which a layer of compacted clay was covered with either 2 ft of topsoil, an unprotected geomembrane, or a geomembrane overlain by 18 in. of topsoil. In less than a year, significant drying and cracking occurred in the plots with soil cover alone and an unprotected geomembrane cover, but no significant desiccation occurred in the clay covered with both a geomembrane and soil. These findings suggest that the best and perhaps only way to protect a compacted clay liner used as a barrier layer in a final cap from desiccation-induced damage is to cover the clay liner with both a geomembrane and a layer of soil.

It appears that few designers have considered the possibility of desiccation from below the clay liner. Compacted clay liners are usually placed at water contents somewhat higher than optimum. If the waste or foundation layer is fairly dry, moisture can be drawn out of the clay and perhaps, in the extreme, cause cracking of the clay. A geomembrane placed below the clay liner can prevent this from occurring, but then the barrier would consist of a triple composite: geomembrane over clay over geomembrane. In addition to added cost, it is probably impractical to compact clay on top of a geomembrane without risk of damage to the geomembrane. Thus, if a triple composite liner is even considered, a geomembrane–geosynthetic clay liner–geomembrane is the most logical choice. No known instances exist in which a triple composite liner has been used for a final cap.

Yet another problem with compacted clays is potential damage from differential settlement. It is convenient to define ''distortion'' as the differential settlement that occurs over a horizontal distance L, or Δ/L. The definition is illustrated in Fig. 5.19. Murphy and Gilbert (1985) computed the relationship between distortion and tensile strain for caps. They also summarized data on the maximum tensile strain that compacted clays can withstand before they crack and found that the maximum tensile strain is about 0.1 to 1%. For these tensile strains, the maximum distortion (Δ/L) that the clay liner can withstand is estimated to be 0.05–0.1.

An example of what this distortion means is as follows. Suppose that a circular depression develops in a cap, and the depression has a diameter of 20 ft. What is the maximum settlement at the center of the depression before the clay liner would theoretically crack? The horizontal distance L is the radius of the circular depression, or 10 ft. The maximum Δ/L value of 0.05–0.1 multiplied by 10 ft. yields a maximum value of 0.5–1 ft. In other words, if the settlement of the center of the depression exceeds 0.5–1 ft., the clay would theoretically crack and lose its low permeability. For most compressible wastes, for example, municipal refuse or mixed drums, this level of distortion or more is probable.

Further confirmation of the damaging effects of differential settlement on the permeability of clay cap materials is provided by Jessberger and Stone (1991), who performed centrifuge tests on 1.4- and 1.8-in.-thick prototype clay liners. A "trap door" beneath the liner was slowly deformed to produce a depression like one would anticipate in a landfill cap. The clay was permeated during the process of deformation, and flow rates were measured. It was found that the flow rate through the clay remained low and steady until the distortion reached $\Delta/L = 0.1$, at which point the rate of flow increased sharply. Tension cracks were noted in the clay when the tests were dismantled. Thus, the experiments reported by Jessberger and Stone (1991) support the findings from Murphy and Gilbert (1985), that is, compacted clay liners in caps cannot withstand a distortion (Δ/L) that is greater than approximately 0.05–0.1 without cracking. In contrast, geosynthetic clay liners can withstand much larger distortions, as discussed in Section 5.3.

Two other types of barrier materials warrant mentioning: waste material used as a barrier (Fig. 5.20) and asphalt (Fig. 5.21). Waste materials that might be used include paper mill waste, which has been studied in field experiments and used for a solid waste landfill cover at a site in western Massachusetts, and combustion ash mixed with bentonite. Laboratory experiments indicate that these, and probably various other types of waste materials, can achieve the common regulatory requirement of a hydraulic conductivity of 10^{-7} cm/s. However, biodegradable materials, such as paper waste, may undergo changes over time, and ash-based materials tend to be brittle and vulnerable to cracking. While waste materials may be attractive materials to use in caps, they could probably not be used in an application involving hazardous waste without further research and development.

Asphalt can be applied as a sprayed-on membrane. Multiple applications are used to ensure complete coverage. Various chemicals can be added to asphalt to improve its properties, or the asphalt could be used without additives. The fact that naturally occurring asphalt is known to have survived for millions

Layer 4 -- Barrier Layer

Alternate F: Waste Material for Barrier Layer

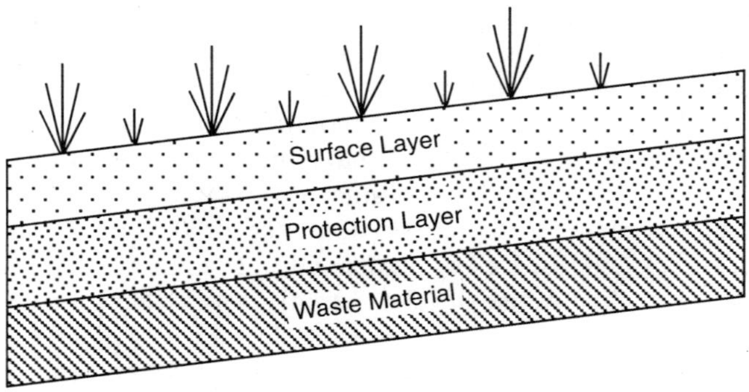

Advantages

1. Waste materials are put to productive use.
2. Some waste materials may have very desirable characteristics and perform very well as a barrier layer.

Disadvantages

1. Experience with use of waste materials in final caps is very limited.
2. Only a few waste materials are likely to be suitable for use as a barrier material.
3. Long-term performance of some waste materials is questionable -- these materials may be more appropriate for interim caps than final caps.

Assessment

Waste materials such as paper mill waste or combustion ash may have low hydraulic conductivity either alone or after treatment with one or more amendments. The concept of utilizing waste materials for construction of caps has strong appeal for some individuals (particularly those who generate the wastes that might be used in a cap), but field-scale data are very limited. Use of wastes shows promise but needs further research and development.

Figure 5.20 Waste material used for barrier layer.

Layer 4 -- Barrier Layer

Alternate G: Asphalt

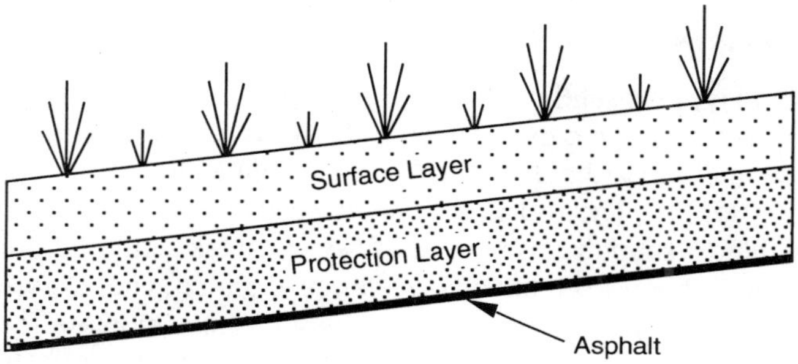

Advantages

1. Straightforward and rapid installation.
2. Materials are readily available.
3. Liner not damaged by wet/dry cycles.
4. If properly protected, asphalt should function for a very long time.
5. Asphalt can self-heal to some extent by cold flowing.
6. Easy to repair.

Disadvantages

1. Ashpalt may be degraded if not properly protected.
2. Vulnerable to puncture.
3. Limited hydraulic effectiveness due to lack of composite action.
4. Type and placement of cover soils is critical to minimize punctures.
5. Very limited experience with asphalt barriers in liners or caps.
6. Asphalt is soluble in some organic chemicals.

Assessment

The main problems with this liner are that asphalt can be punctured during construction and later from accidental intrusion, and asphalt may degrade if not buried at a suitable depth and protected properly.

Figure 5.21 Asphalt used for barrier layer.

of years is an advantage and it is generally thought that if asphalt in a cap is suitably protected from ultraviolet (UV) light and oxidation, it too will have an indefinite life. There is practically no prior experience with asphalt used in final caps, but it is being considered for caps at some DOE facilities where the cap must last for at least 10,000 years.

5.2.5 Foundation and/or Gas Collection Layer

A foundation layer is usually needed for the barrier layer. The foundation layer may or may not serve as a gas collection layer, too. The materials that can be used for foundation (without gas collection) include soil (Fig. 5.22) or waste material (Fig. 5.23). The foundation layer needs to be reasonably strong and incompressible to support the weight of heavy construction equipment. Practically any type of soil (except for wet clay or organic material) will serve this purpose, and many types of waste materials will make suitable foundations.

Natural or geosynthetic materials can be used for a gas collection layer, if a gas collection layer is needed. Sand or gravel (Fig. 5.24) is probably the most common material. A filter may be needed to separate the sand or gravel from an overlying barrier layer, depending on the materials involved. The granular materials need to be relatively dry in order to maintain a high permeability to gas.

A geotextile (Fig. 5.25) or geonet (Fig. 5.26) can also be used to collect landfill gases. A foundation layer will probably be needed beneath these materials. A potential advantage of thin geosynthetic materials is that the weight of these materials is very low, which is advantageous when compressible waste or soil underlies the cap. Care must be exercised to ensure that the geotextile has adequate transmissivity under realistic field conditions.

5.3 GEOSYNTHETIC CLAY LINERS FOR BARRIER LAYER IN FINAL CAPS

Geosynthetic clay liners represent a new, alternative material that deserves serious consideration for use in final caps. In this section, data (much of it unpublished) will be presented to illustrate the engineering characteristics of GCLs with emphasis on performance criteria, such as hydraulic properties, resistance to wet/dry cycles, and impact of differential settlement.

5.3.1 Types of Geosynthetic Clay Liners

Geosynthetic clay liners fall into two broad categories: (1) GCLs with bentonite sandwiched between two geotextiles, and (2) bentonite mixed with an adhesive and glued to a geomembrane. Within these two broad categories, four types of geosynthetic clay liners are currently manufactured (Fig. 5.27):

Layer 5 -- Foundation or Gas Collection Layer

Alternate A: Soil Serving as Foundation for Barrier Layer

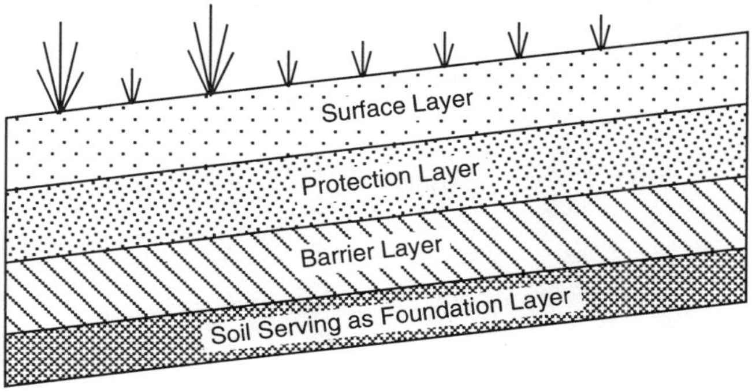

Advantages	Disadvantages
1. Long history of use of soil as foundation material in engineering works. 2. Almost any locally available soil is likely to make a suitable foundation for the barrier layer. 3. Material requirements are very straightforward if this layer does not have to serve as a gas collection layer, also.	1. If gas control is required, the foundation layer will also have to serve as a drainage layer, which leads to many more design requirements and constraints on suitable materials.

Assessment

The use of soil as a foundation material is an obvious choice, particularly for sites where soil material is readily available.

Figure 5.22 Soil used for foundation layer.

Layer 5 -- Foundation or Gas Collection Layer

Alternate B: Waste Material Serving as Foundation for Barrier Layer

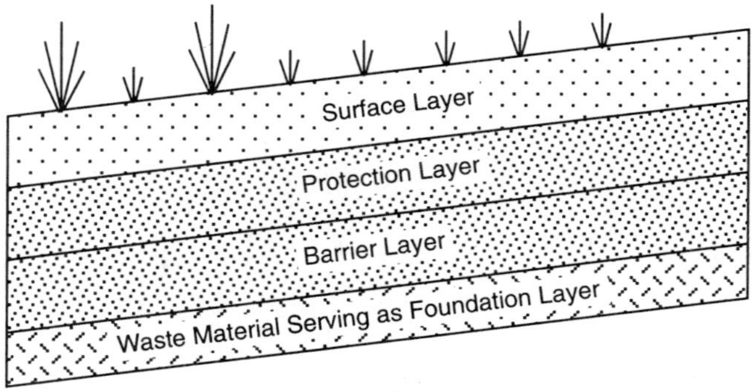

Advantages

1. Leads to productive use of materials that would otherwise be considered wastes.
2. Some waste materials can make an excellent foundation.

Disadvantages

1. If gas control is required, the foundation layer will also have to serve as a drainage layer , which leads to many more design requirements and constraints on suitable materials.
2. Not all waste materials are suitable for use as a foundation layer.
3. If materials are not available nearby, transportation costs could be very large.

Assessment

The use of waste material as a foundation is a reasonable choice, provided suitable waste materials are available reasonably close to the disposal site.

Figure 5.23 Waste material used for foundation layer.

Layer 5 -- Foundation or Gas Collection Layer

Alternate C: Sand or Gravel Drain with Soil Filter

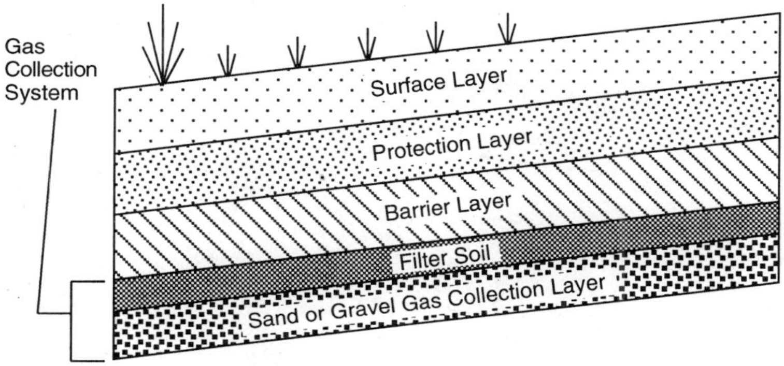

Advantages

1. Long history of use of soil filters and sand or gravel drains in engineering works.
2. Easy to install perforated collection pipes in sand or gravel gas collection layer.
3. Base of vent pipes are easy to locate between filter and bottom of gas collection layer.

Disadvantages

1. Suitable materials may not be readily available locally.

Assessment

This is a technically sound design and probably the most common design when suitable construction materials are available locally.

Figure 5.24 Sand or gravel used for gas collection layer.

Layer 5 -- Foundation or Gas Collection Layer

Alternate D: Geotextile Drain

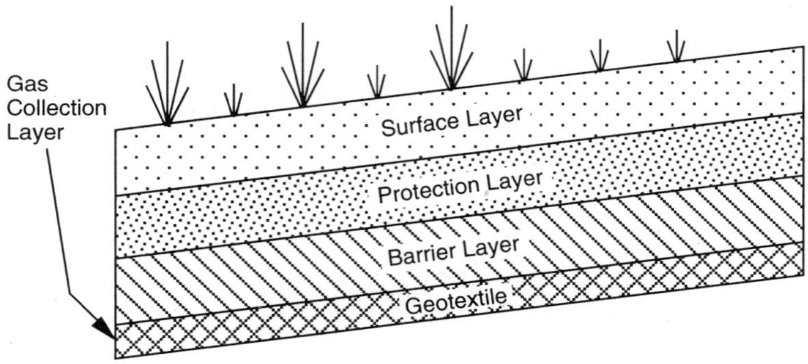

Advantages

1. Geotextiles are readily available and can be shipped to any location.
2. Provides clean working surface if geosynthetics are used as lower portion of barrier layer.
3. Rapid, straightforward installation.
4. Specialty geotextiles can be fabricated for site-specific designs, e.g., high density fibers and composite geotextiles needled together.
5. Weight of material is very small, which minimizes settlement.

Disadvantages

1. Potential slippage between geotextile and overlying geomembrane (if the barrier layer has a geomembrane on the lower surface of the layer).

Assessment

This design is considered technically equal to designs that employ sand filter and gravel drainage materials. Selection of natural materials or geosynthetics is often based upon local availability and cost of natural materials and construction logistics. In many cases, this would be the least costly of the gas collection layer alternatives.

Figure 5.25 Geotextile used for gas collection layer.

Layer 5 -- Foundation or Gas Collection Layer

Alternate E: Geonet Drainage Material with Geotextile Filters

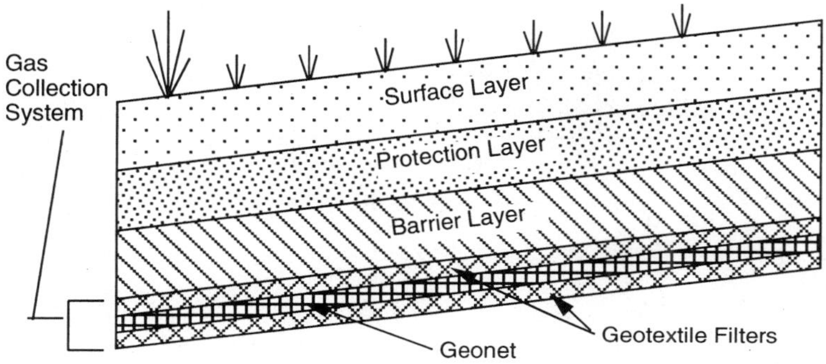

Advantages

1. Geotextiles and geonet are readily available and can be shipped to any location.
2. Geonet has extremely high in-plane gas transmissivity.
3. Geotextiles can be bonded to the geonet using a variety of fabrication methods.
4. Provides clean working surface if geosynthetics are used as lower portion of barrier layer.
5. Weight of materials is very small, which minimizes settlement of underlying waste or soil.

Disadvantages

1. Potential slippage between geotextile and overlying geomembrane (if the barrier layer has a geomembrane on the lower surface of the layer) and between geonet and geotextiles.
2. Continuity of flow within the geonet must be assured across seams and edges.
3. Somewhat more complex connection to vent pipes than a single, thick geotextile drain and filter.

Assessment

This design is considered technically equal to designs that employ sand filter and gravel drainage materials. Selection of natural materials or geosynthetics is often based upon local availability and cost of natural materials and construction logistics. The geonet has a higher in-plane gas transmissivity than a geotextile, which could be advantageous if gas production rates are extremely large or vent pipes are more widely spaced than usual.

Figure 5.26 Geonet used for gas collection layer.

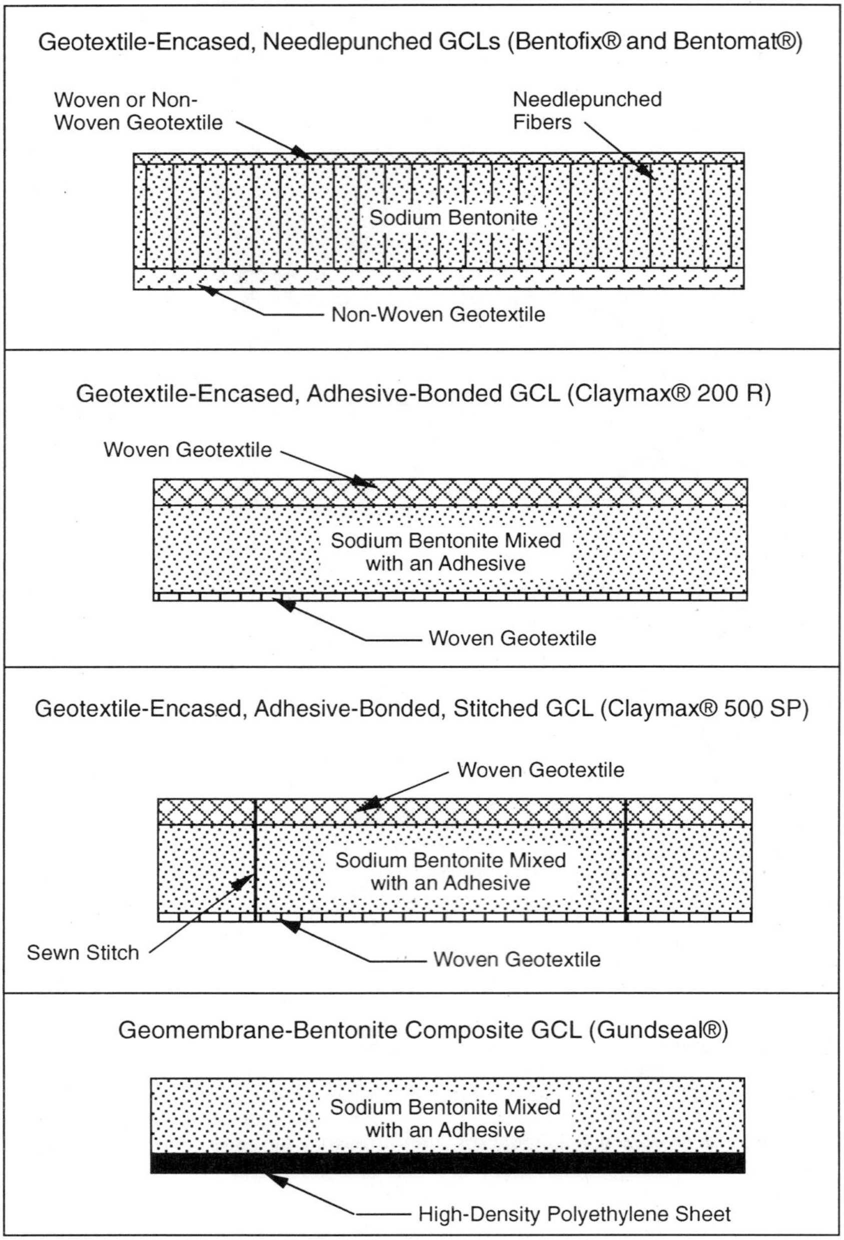

Figure 5.27 Four types of geosynthetic clay liners.

- Geotextile-encased, needlepunched GCLs (Bentofix® and Bentomat®):
- Geotextile-encased, adhesive-bonded GCLs (Claymax® 200 R):
- Geotextile-encased, adhesive-bonded, stitched GCLs (Claymax® 500SP):
- Geomembrane-bentonite composite GCL (Gundseal®).

In the first group, Bentofix® and Bentomat® consist of bentonite sandwiched between a woven and nonwoven geotextile that are needlepunched together. Claymax® 200R consists of bentonite mixed with a water-soluble glue and sandwiched between two woven geotextiles; Claymax® 500SP is the same except that the two woven geotextiles are sewn together to provide added internal shear strength. Gundseal® consists of a special adhesive that will adhere to high-density polyethylene (HDPE) that is mixed with bentonite, and is glued to an HDPE or very low density polyethylene (VLDPE) geomembrane.

All the GCLs contain approximately 1 psf of sodium bentonite. The materials have a dry thickness of about $\frac{1}{4}$ in. In a cover application, the thickness after hydration would typically be $\frac{1}{2}$ in. The GCLs are manufactured in panels with widths of approximately 13–17 ft and lengths of 75–200 ft. The panels are placed on rolls at the factory and are unrolled at the time of installation. The panels are overlapped 6–12 in. during installation and are said to be "self-sealing" at the overlap. A sketch of the overlap zones is shown in Fig. 5.28. With the geotextile-encased, needlepunched GCLs, additional sodium bentonite is placed along the overlap at a rate of approximately 0.25 psf. The bentonite penetrates the pores of the geotextiles and is said by the manufacturer to cause the material to self-seam when the bentonite hydrates. With Claymax® the manufacturer states that the bentonite oozes out through the openings in the geotextiles and causes the material to self-seal without the need for any additional bentonite along the overlap. With Gundseal® the GCL can be placed with the bentonite facing upward (Fig. 5.27) or (as shown in Fig. 5.28) downward. The material is said to be self-sealing at the overlap.

Although no mechanical joining of overlapped seams is necessary for any of the GCLs, the geomembrane component of Gundseal® can be welded together along the overlaps, if desired, but a thicker sheet than the usual 20 mil sheet would be required. Gundseal® would be placed with the bentonite facing upward if a separate, conventional geomembrane lining is to be placed on top of the GCL.

5.3.2 Engineering Properties

a. Hydraulic Conductivity. Hydraulic conductivity data on the GCL materials are summarized in Fig. 5.29. Data on the Gundseal® are for the bentonite component only. The tests have been performed by various organizations over a range in confining stress. The hydraulic conductivity of GCLs varies between about 10^{-10} and 10^{-8} cm/s.

For a final cap, a confining stress on the order of 10–30 kPa is a reasonable

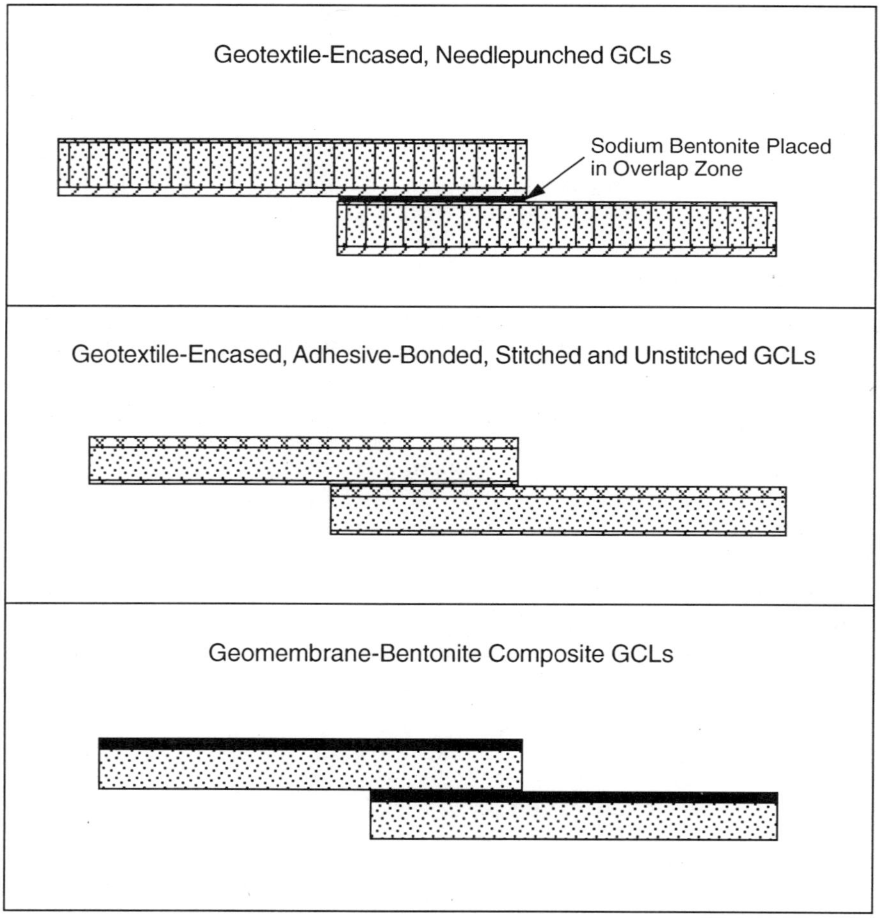

Figure 5.28 Overlapped seams in geosynthetic clay liners.

range. There are some differences between the hydraulic conductivities of the various GCLs, but the differences do not appear to be very large (except for the geomembrane–bentonite composite GCLs, which have a much lower hydraulic conductivity (equivalent of 10^{-12} cm/s) if the geomembrane component is included in the determination of hydraulic conductivity and provided there are no holes in the membrane and the overlap seal is fully effective). For a final cap, it is reasonable to assume that several, if not all, GCLs can be supplied with a hydraulic conductivity of 1×10^{-9} cm/s if properly placed without damage.

This hydraulic conductivity is well below most regulatory standards of 1×10^{-7} cm/s, although the thickness of the GCL is much less that the typical thickness of 1–3 ft for a compacted clay liner. Hydraulic equivalency of a GCL

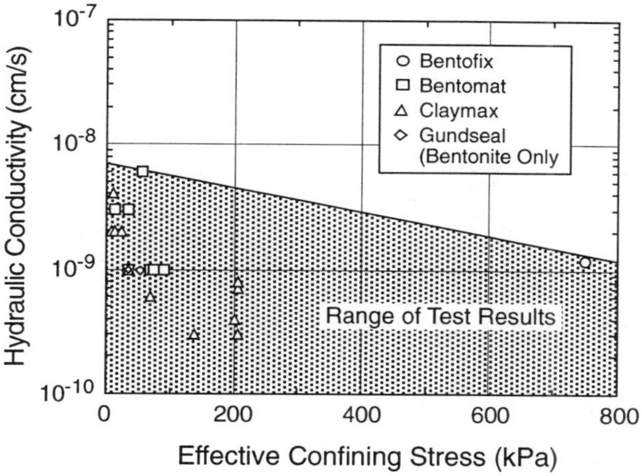

Figure 5.29 Hydraulic conductivity of geosynthetic clay liners.

to a compacted clay liner may be evaluated from Darcy's law, which states that

$$q = k\left(\frac{H + L}{L}\right)A \tag{1}$$

where q is the rate of flow, k is hydraulic conductivity, H is the head of liquid ponded on a liner of thickness L, and A is the cross-sectional area of flow. If a GCL is equivalent to a CCL, then the flow rates through each liner would be equal:

$$k_{\mathrm{GCL}}\left(\frac{H + L_{\mathrm{GCL}}}{L_{\mathrm{GCL}}}\right)A = k_{\mathrm{CCL}}\left(\frac{H + L_{\mathrm{CCL}}}{L_{\mathrm{CCL}}}\right)A \tag{2}$$

The required hydraulic conductivity of the GCL to establish equivalent performance to a compacted clay liner may be computed by rearranging Eq. 2 as follows:

$$k_{\mathrm{GCL}} = k_{\mathrm{CCL}}\left(\frac{L_{\mathrm{GCL}}(H + L_{\mathrm{CCL}})}{L_{\mathrm{CCL}}(H + L_{\mathrm{GCL}})}\right) \tag{3}$$

For example, Eq. 3 may be applied for the following typical conditions: $k_{\mathrm{CCL}} = 1 \times 10^{-7}$ cm/s, $L_{\mathrm{GCL}} = 1$ cm, $L_{\mathrm{CCL}} = 60$ cm (2 ft), and $H = 30$ cm (1 ft.). For these assumptions, $k_{\mathrm{GCL}} = 5 \times 10^{-9}$ cm/s. If equivalency is borderline, or if added redundancy is desired, two layers of GCLs can be used instead of one.

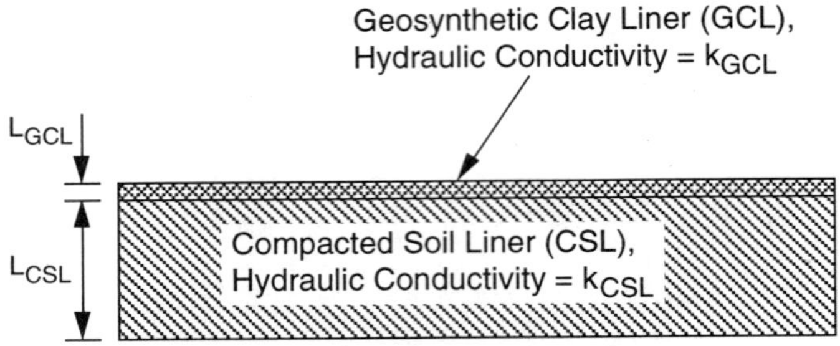

Figure 5.30 Definition of terms for hydraulic equivalency analysis.

A GCL can also be used in conjunction with a layer of compacted soil as shown in Fig. 5.30. In such cases, the compacted soil will tend to be thinner or be of higher hydraulic conductivity compared to the minimum requirements for compacted clay liners usually stated by regulatory agencies. If the compacted soil liner were neither thinner nor more permeable than required by regulation, there would be no motivation to use a GCL, other than to provide redundancy. By employing a GCL and a compacted soil liner (CSL) of hydraulic conductivity k_{CSL}, that is greater than the usual requirement for a compacted clay liner, one may be able to achieve an acceptable alternative to a conventional compacted clay liner. The equivalent hydraulic conductivity of the composite GCL–CSL may be computed from the following equation:

$$k_{eq} = \frac{L_{GCL} + L_{CSL}}{\dfrac{L_{GCL}}{k_{GCL}} + \dfrac{L_{CSL}}{k_{CSL}}} \tag{4}$$

One can equate the flow rate through the GCL–CSL composite with that through a conventional compacted clay liner and determine the hydraulic conductivity of the GCL (k_{GCL}) needed to provide an equivalent liner from the following equation:

$$k_{GCL} = \frac{L_{CCL}}{k_{CCL}}\left(\frac{H + L_{CSL}}{H + L_{CCL}}\right) - \frac{L_{CSL}}{k_{CSL}} \tag{5}$$

For example, if the compacted clay liner has $k_{CCL} = 1 \times 10^{-7}$ cm/s and $L_{CCL} = 60$ cm (2 ft), and the compacted soil liner has $k_{CSL} = 1 \times 10^{-6}$ cm/s and $L_{CSL} = 30$ cm (1 ft.), then, for a 1-cm-thick GCL and $H = 30$ cm, the required hydraulic conductivity of the GCL is 2.7×10^{-9} cm/s. This value is practically identical to the required value for the GCL to exhibit equivalency by itself—the 1-ft-thick layer of soil having a hydraulic conductivity of 1×10^{-6} cm/s adds little to the theoretical hydraulic effectiveness of the composite. However, the idea of combining GCLs with native soils is very appealing to some individuals because of the redundancy that the combination provides. The situation depicted in Fig. 5.30 and described in this paragraph is presented primarily to illustrate the options available to the designer in trying to meet regulatory agency concerns by using nonstandard materials or designs.

The preceding discussion has dealt entirely with the hydraulic performance of the parent GCL material. However, overlaps are of equal, if not greater, concern. Estornell and Daniel (1992) tested the behavior of overlapped panels of GCL materials in steel tanks that measured 8 ft in length, 4 ft in width, and 3 ft in height. At least one test was performed using the minimum overlap width recommended by the manufacturer, and at least one test was conducted using one half the minimum recommended overlap width. Results indicate that the overlapped seams of the GCLs tested performed very well. Flow rates through control specimens that contained an overlap were the same as flow rates through the parent materials. Conclusions drawn from this study included the following:

1. The GCLs self-sealed at overlaps when the bentonite was hydrated. The hydraulic conductivities of overlapped materials were almost identical to values measured on control samples with no overlap.
2. In most cases, the overlap width could be reduced up to 50% below the minimum values recommended by the manufacturers without significantly increasing the overall hydraulic conductivity of overlapped panels. However, reducing the overlap is not recommended because it would pose difficulties for control in the field. For the carefully controlled test conditions employed in this investigation, there was a factor of safety built into the manufacturer's minimum recommended overlap widths.

Shan and Daniel (1991) studied the effects of punctures, wet–dry cycles, and freeze–thaw cycles on one GCL (Claymax® 200R). Results were as follows:

Punctures. The effects of punctures on the hydraulic conductivity of the GCL were studied by drilling or cutting circular holes into the dry GCLs, setting the punctured GCL up in a flexible-wall permeameter, and permeating the GCL slowly until steady flow was achieved. Results are summarized in Table 5.1.

Small (≤ 1 in.) punctures made in the dry material were fully sealed by the highly expansive bentonite upon hydration of the bentonite. These tests demonstrate the high self-healing capability of bentonite.

Wet–Dry Cycles. Samples of the GCL were permeated in a flexible-wall permeameter, removed from the permeameter, and allowed to air dry under a small applied vertical stress. All specimens underwent severe cracking upon drying. The specimens were then set back up in a flexible-wall permeameter,

TABLE 5.1 Variation of Hydraulic Conductivity with Puncture Diameter

Diameter of Puncture (in.)	Hydraulic Conductivity (cm/s)
No Punctures	2×10^{-9}
0.5	3×10^{-9}
1.0	5×10^{-9}
3.0	$>1 \times 10^{-4}$

slowly rehydrated, and then repermeated. There was no change in hydraulic conductivity from the initial value of 2×10^{-9} cm/s, even after three wet–dry cycles. These tests reinforce the fully reversible shrink–swell nature of bentonite.

Freeze–Thaw Cycles. The hydraulic conductivity of a GCL specimen was determined. Then the specimen was removed from the flexible-wall permeameter and subjected to five freeze–thaw cycles. The specimen was repermeated, and it was found that the hydraulic conductivity did not change.

The manufacturers of GCLs have had similar tests performed by independent laboratories to evaluate self-healing characteristics of their particular products. Results are similar to those reported by Shan and Daniel (1991) for Claymax® 200R.

b. Shear Strength. The shear strength of Bentomat,® Claymax® 200R, and Gundseal® has been determined by Daniel (1993b). The manufacturers of the GCLs have had shearing tests performed on their materials by independent testing laboratories, but in most cases the rates of shear are thought to be too high to yield conservative values and therefore are not discussed in this chapter. The following results are for conditions of full saturation of the GCL with water and shearing under consolidated-drained (extremely slow) conditions. Direct shear boxes were used to perform the tests. Test specimens were cut from parent material, set up in a direct shear apparatus, and subjected to the desired normal load. Then the specimens were soaked with water and allowed to equilibrate; about 3 weeks were required before swelling ceased. The specimens were then sheared very slowly with failure occurring in 3–7 days. Results are summarized in Table 5.2.

Careful examination of the low-normal-stress region shows that the failure envelope is actually curved. This has significant impact for the stability of final caps, which, due to their thinness, exert low normal stress on the soil. In studies underway at the University of Texas, tilt table tests are being performed. Samples of GCL materials that measure 12 in. × 12 in. are set up on a tilt table, loaded with a steel plate, placed in a water bath, and allowed to

TABLE 5.2 Effective Cohesion and Angle of Internal Friction for Various Geosynthetic Clay Liners

Geosynthetic Clay Liner	Effective Cohesion (psi)	Angle of Internal Friction (degrees)
Bentomat®	4.4	29
Claymax®	0.6	9
Gundseal®	1.2	8

fully hydrate. Then the table is slowly tilted over a period of several weeks until sliding occurs.

The direct shear and tilt table tests indicate that the failure envelope is curved. Figure 5.31 presents the relationship between angle of internal friction and normal stress. Clearly, the angle of internal friction is sensitive to normal stress, and the angle of internal friction is greatest at low normal stresses, such as those expected in final caps. For final caps, a typical range of normal stress is approximately 200–700 psf.

There recently have been data generated, but as-yet unpublished, to suggest that some of the internally reinforced GCLs (e.g., needle punched materials) can exhibit a substantial drop-off in strength when these materials are sheared due to horizontal deformations in excess of 0.1–1.0 in. With deformation, the needle punched fibers or other reinforcing elements tend to either break or pull out, causing the shear strength at large deformation (so-called ''residual shear strength'') to be substantially less than the peak value. The designer should carefully consider how much deformation is possible and then use strength

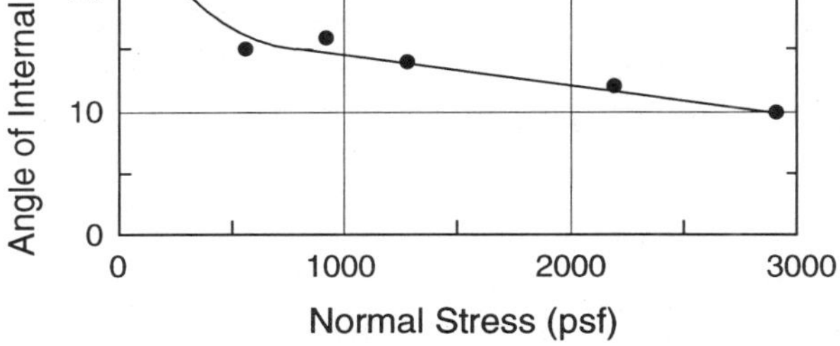

Figure 5.31 Relationship between angle of internal friction and normal stress for water-saturated Bentonite in one GCL (Gundseal®).

parameters that are consistent with the anticipated deformation. Manufacturers of GCLs are working to strengthen the reinforcement in an effort to control this potential problem.

The shear strength of the interface between a GCL and adjacent components should also be considered. Extrusion of bentonite through the openings of geotextiles, for GCLs with geotextile-encased bentonite, can lead to lubrication of the interface and a reduction in interfacial shear strength. Tests to determine interfacial strengths are recommended on a site-specific basis.

Finally, it is possible to sandwich the layer of bentonite between two geomembranes. This is a particularly attractive alternative in the Gundseal®: the geomembrane side of the GCL would face downward and a separate geomembrane would be installed on top of the bentonite layer. The only mechanisms by which the bentonite could become wet are: (1) molecular diffusion of water through one or both geomembranes; (2) suction of water into the bentonite along overlaps of the GCL panels (where the bentonite is exposed); and (3) migration of water through defects in the geomembrane or seams in the geomembrane. With good construction quality assurance, defects in the geomembrane or its seams should be minor and infrequent (one or two very tiny imperfections per acre, primarily in seams, might be a typical expectation). The width of Gundseal® panels (17 ft) causes overlaps to be widely spaced. Molecular diffusion occurs so slowly that, according to calculations based on published rates of water vapor transmissions through HDPE geomembranes, significant wetting of the bentonite should take many centuries or millennia. Thus, a viable design alternative is to sandwich the bentonite between two geomembranes and to assume that the bentonite remains dry and, hence, relatively strong. This approach is only appropriate when careful design and rigorous construction quality assurance guarantee that there will not be numerous defects in the geomembranes.

c. Response to Differential Settlement. LaGatta (1992) studied the effects of differential settlement on the hydraulic performance of GCLs. The same tanks that Estornell and Daniel (1992) used were modified by LaGatta, who placed a water-filled bladder beneath the GCLs and incrementally deflated the bladder below the GCL to produce differential settlement to model what would happen in a landfill cap. LaGatta measured the relationship between deformation and hydraulic conductivity for GCLs with and without an overlap. Results for one GCL (Bentomat®) are summarized in Table 5.3.

The extreme differential settlement caused by the deflated bladder did not produce an increase in hydraulic conductivity for Bentomat®. Somewhat similar results have been obtained for Gundseal® and Claymax® 500SP; Bentofix® has not been tested. Thus, it appears that many GCLs can withstand large deformation (Δ/L up to 0.5) and tensile strain (up to 10–15%) without undergoing large increases in hydraulic conductivity.

LaGatta (1992) summarized data, presented in Table 5.4, on the tensile

TABLE 5.3 Deformation, Tensile Strain, and Hydraulic Conductivity as a Function of Overlap Condition

Overlap Condition	Deformation Δ/L	Tensile Strain (%)	Hydraulic Conductivity (cm/s)
Intact specimen	0.347	6.0	1×10^{-9}
Overlapped panels	0.504	12	1×10^{-9}

strain at failure for compacted clay. From this table it is seen that normal soils cannot withstand tensile strains greater than approximately 0.85% without failing (cracking). Bentonite, on the other hand, is reported by Tschebotarioff and DePhilippe (1953) to have a tensile strain at failure of 3.4%, but LaGatta measured much greater tensile strains with no cracking in the bentonite component of GCLs, probably due to the beneficial reinforcing effects of the geotextiles or geomembrane in the GCLs, as well as higher water content. In any case, bentonite can obviously withstand much greater tensile deformation than normal compacted soils without cracking, which is a very favorable characteristic for final caps.

TABLE 5.4 Tensile Strain at Failure for Various Compacted Clays

Reference	Type or Source of Soil	Water Content (%)	Plasticity Index (%)	Failure Tensile Strain
Tschebotarioff and DePhilippe (1953)	Natural clayey soil	19.9	7	0.80%
	Bentonite (montmorillonite)	101	487	3.4%
	Illite	31.5	34	0.84%
	Kaolinite	37.6	38	0.16%
Leonards and Narain (1963)	Portland Dam Portland, CO	16.3	8	0.14%
	Rector Creek Dam Napa, CA	19.8	16	0.16%
	Woodcrest Dam Riverside, Ca	10.2	Nonplastic	0.18%
	Shell Oil Dam Ventura, CA	11.2	Nonplastic	0.07%
	Willard Test Dam Embankment, UT	16.4	11	0.20%

5.4 INTERIM CAPS

The previous sections have focused on final caps. Often, an interim cap is desired while appropriate options are considered. The purpose of an interim cap is usually to provide source control by minimizing infiltration of water and leaching or transport of waste constituents. There may be other objectives for a specific project that may include control of vapors, redirecting surface runoff, and separation of the waste from plants and animals. The only fundamental difference between an interim cap and a final cap is that the design life of an interim cap is limited, while that of a final cap is essentially unlimited. Therefore, materials that have a finite operational life (e.g., some geosynthetic materials) may be very attractive materials for an interim cap but would not be considered for a final cap.

The five components of a cap shown in Fig. 5.1, and the materials discussed previously, are used for interim caps. As with final caps, all five of the potential components are not needed for all interim caps—just those that are needed should be included in the design. For example, an internal drainage layer above the barrier layer is rarely employed with interim caps. Also, geomembranes are extensively used as the barrier layer in interim caps due to the high cost effectiveness of geomembranes and the limited lifetime of the interim cap. Except for the special considerations just noted, the design considerations for an interim cap are essentially the same as those of a final cap.

5.5 ASSESSMENT FOR OVERALL CONTAINMENT

Unfortunately, there are practically no publications that describe the actual performance of caps. One reason is that the performance is difficult to quantify—one cannot readily determine exactly how much water is percolating through a cap or how much gas might be escaping. An additional reason for such meager information of performance of caps is that the modern multicomponent cap has only been in existence for about a decade, and too little time has passed to enable much performance information to be gathered. Finally, caps are often used in conjunction with pump-and-treat remedies, and it is difficult to determine how much of the water entering the pump-and-treat system is from infiltration of water through the cap.

The comments that are made in this section are a reflection of general experience, direct observation of caps, and discussions with facility owners or their consultants who have designed or maintained caps. Experience with caps has been mixed and depends significantly on the following factors:

- Slope angle of the final cap.
- Annual precipitation.

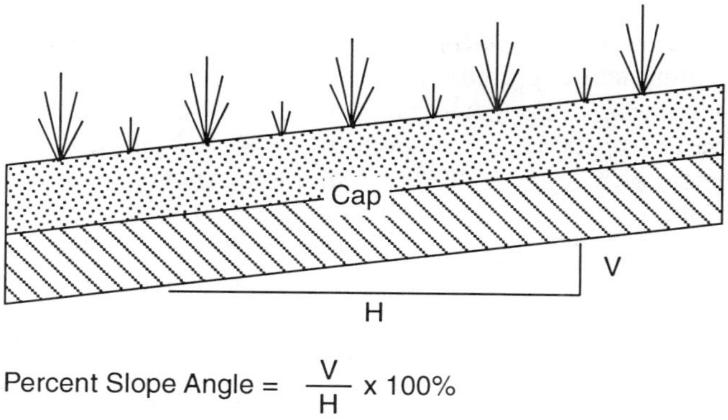

Figure 5.32 Definition of percent slope in caps.

- Presence of burrowing animals.
- Components included in the cap design, thickness of the components, and the materials used for the components.
- Compressibility of the underlying waste.
- Gas generation from the waste.

5.5.1 Slope Angle

Caps must have a minimum slope angle to promote runoff of precipitation. The percent slope angle is defined in Fig. 5.32. A typical minimum slope angle is 5%. Often, the maximum slope angle is considerably greater than 5%; slopes inclined as steeply as 100% (a 45° angle) have been employed, but slope inclinations rarely exceed 33–40%. As shown in Fig. 5.33, the interior part of a cap may be relatively gently sloping, but the edges of the cap tend to be more steeply sloping.

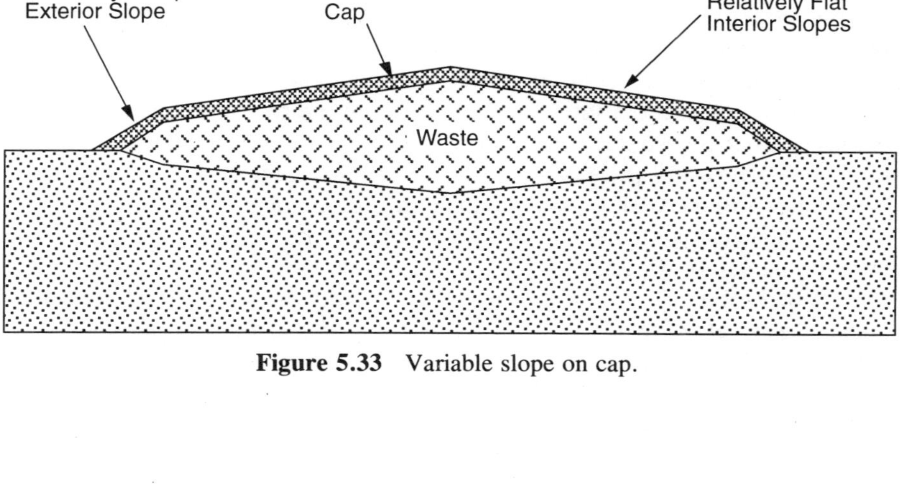

Figure 5.33 Variable slope on cap.

The problem posed by slope inclination is as follows. If the slope angle is near the minimum angle normally employed (5%), any settlement of the underlying waste can alter drainage patterns and create closed drainage areas where water may pond. If the slope is too steep, problems with erosion and development of rills and gullies are common. Altered surface drainage on relatively flat covers and excessive erosion on relatively steep covers, are probably the two most common problems that can be readily observed with existing caps.

5.5.2 Annual Precipitation

Caps in humid climates tend to be challenged much more severely in terms of limiting infiltration of water. In arid climates, evapotranspiration may equal or exceed precipitation, which means that there is little or no percolation of water through a cap even for an unengineered cap consisting of soil with vegetation planted on the soil. In humid climates, the soils tend to be soaked with water frequently, which not only promotes percolation of water through the cap but can also create problems with slope instability.

Regions with more than about 15 in. of precipitation per year pose greater challenges in terms of stability and limiting percolation of water through the cap than regions with less precipitation. For arid regions (15 in. per year of precipitation), the main problem tends to be animal burrows through the cap, and development of erosion gullies on sparsely vegetated caps. Thus, arid and humid regions pose fundamentally different challenges in terms of cap design.

5.5.3 Burrowing Animals

As stated in the previous section, burrowing animals can be a major problem at arid sites. Not only can the burrowing animals dig into the waste and consume the waste directly, but the burrow holes provide access for other animals and insects, as well. Serious interest in "biobarriers" within a cap to stop burrowing animals has been focused almost entirely upon problems with disposal of radioactive waste at arid sites such as Los Alamos, Idaho Falls, and Hanford. Radioactive materials can be consumed by certain species and, by bioaccumulation, pose a threat at some point in the food chain other than the point of initial ingestion or inhalation. The problem of burrowing animals is rarely considered in cap design, except for the three radioactive waste disposal sites just mentioned. Burrowing animals are of concern only for some relatively arid sites, and methods are available to discourage animal burrows through the cap (e.g., placement of a layer of cobbles in the cap as shown in Fig. 5.8). Problems with burrowing animals have been experienced in unengineered caps at Hanford, WA, but there are few documented problems with animal burrows except for the Hanford site.

5.5.4 Components of a Cap

Many problems with caps over landfills have been observed due to inadequate material selection or thickness of materials in the cap. The most common problems are

- Lack of adequate gas collection, leading to uncontrolled release of gas or build up of gas pressure that contributes to instability of the cap or buried waste.
- Inadequate foundation for the cap, leading to collapse of the cap into open cavities.
- Use of a compacted clay barrier layer in situations where the clay is doomed to fail to perform as designed because of large differential settlement or inadequate protection from freeze–thaw or wet–dry cycles.
- Slippage of soil along a soil-geomembrane interface resulting from inadequate interfacial friction between the geomembrane and soil.
- Lack of adequate drainage in the cap, leading to saturation of the cap soils that triggers other problems such as excessive erosion from runoff and instability within the soil cover.
- Lack of adequate soil depth in the protection layer and/or surface layer, leading to exposure of the underlying components to freeze–thaw, wet–dry, or direct exposure of geosynthetics to UV radiation.
- Puncture of geomembranes due to lack of control over the maximum particle size in adjacent soils.
- Use of a topsoil that was too vulnerable to erosion, particularly before growth of grass on the soil had been fully established.

5.5.5 Compressibility of Underlying Waste

Many designers of caps fail to recognize the problems of trying to construct a permanent cap on top of highly compressible waste or soil. Inevitably, the weight of the cap will cause settlement. For heterogeneous soils or wastes, large differential settlement can occur, which causes development of closed depressions in the surface, large cracks, and other problems. If the waste is highly compressible, consideration should be given to stabilizing the waste (in place, if possible), surcharging the waste to force the settlement to occur before the cap is built, building a light-weight cap (for instance, utilizing a geomembrane and geosynthetic clay liner, along with other geosynthetic components wherever possible), or placing an interim cap over the waste until much of the settlement has occurred.

5.5.6 Control of Gas

A common problem with wastes that produce gas is inadequate control of gas. The best way to deal with gas is to construct a continuous gas collection layer at the base of the cap (Fig. 5.1). Vent pipes (typical spacing of 1 pipe per acre) provide a means for collecting the gas and venting it to the atmosphere or pumping the gas to a treatment or energy conversion facility.

5.6 CONSTRUCTION QUALITY CONTROL

As discussed in the preceding sections, the selection of the particular cap design and associated components is based on many considerations such as cost, effectiveness, availability, and stability. An additional consideration is the ability to measure the properties and performance of the constructed product, that is, the ability to implement a meaningful CQC program. The compatibility of cap barrier materials with waste materials is not usually tested since the cap does not usually come in direct contact with waste or leachate, unlike vertical barriers or floors.

5.6.1 Compacted Soil

a. Preconstruction. Compacted soil is a common choice of material for constructing a cap over a waste site in regions where an ample supply of suitable soil is economically available. The CQC program for a compacted soil cap begins when a borrow source that has an adequate supply of suitable soil is identified.

The CQC plan is generally prepared and included with the contract bidding documents to open lines of communication between the involved parties prior to construction. The plan prescribes the CQC organization, that is, the responsibilities of all members of the CQC team and their respective authorities, the methods for performing the CQC inspections, including specifications for performing the tests (both field and laboratory) and the methods and frequency for sampling, as well as equipment calibration requirements. The criteria to be applied in accepting the constructed product are also presented, as well as methods that can be used to correct a product that is not in compliance with design specifications. The reporting requirements (documentation) are also described.

Exploration at the proposed borrow soil location is done before construction to characterize the subsurface conditions (Goldman et al., 1990). Test pits are useful for this exploration. The exploration program should identify the subsurface profile, the engineering properties of soils being considered, the volume of suitable soils, water levels, and the presence of undesirable materials such as organics, contamination, and sand. Samples of the borrow soil are collected and laboratory tested for index properties, such as moisture content (ASTM D2216), Atterberg Limits (ASTM D4318), and particle size distribution (ASTM D422). These tests are used to screen the borrow pit soils and

TABLE 5.5 Recommended Materials Tests for Barrier Layers

Parameter	Test Method	Minimum Testing Frequency
Percent fines[a]	ASTM D1140	1 per 1000 yd^3 [b,c]
Percent gravel[d]	ASTM D422	1 per 1000 yd^3 [b,c]
Liquid and plastic limits	ASTM D4318	1 per 1000 yd^3 [b,c]
Water content	ASTM D4643[e]	1 per 200 yd^3 [b,c,f]
Water content[g]	ASTM D2216	1 per 1000 yd^3 [c,g]
Construction oversight	Observation	Continuous in borrow pit on major projects; continuous in placement area on smaller projects

[a] Percent fines is defined as percent passing the No. 200 sieve.
[b] In addition, at least one test should be performed each day that soil is excavated or placed, and additional tests should be performed on any suspect material observed by QA personnel.
[c] 1000 yd^3 = 836 m^3.
[d] Percent gravel is defined as percent retained on the No. 4 sieve.
[e] This is a microwave oven drying method. Other methods may be used, if more appropriate. Any method used besides direct drying via ASTM D2216 should be calibrated against ASTM D2216 for the onsite solids.
[f] 200 yd^3 = 167 m^3.
[g] Conventional oven drying (ASTM D2216) is recommended on every fifth sample taken for rapid measurement. The intent is to document any systematic error in rapid water content measurement.

check that the borrow index properties are within the specified range. The frequency of performing these tests depends on the soil variability (Goldman et al., 1990). However, the USEPA (1991a) suggests a recommended testing frequency as indicated in Table 5.5.

Samples collected from the proposed borrow site are tested in the laboratory to identify the Acceptance Zone, that is, the combination of moisture content and dry density combinations that result in an acceptable permeability. This usually involves testing the moisture–density relationship of soil samples using three different compactive efforts and measuring associated permeabilities in a laboratory sample (Daniel and Benson, 1990). The compactive efforts are described in the modified Proctor test (ASTM D1557), the standard Proctor test (ASTM D698), and the reduced Proctor effort, which is less than that used in the standard Proctor test. This procedure is followed so that the field permeability can be inferred from measurements of the in-place moisture content and dry density. It assumes that differences in the compaction methods used in the laboratory and in the field do not influence the measured values of permeability. In fact they do, as shown by Mitchell et al. (1965). Fortunately, however, impact laboratory compaction generally yields higher values of permeability than are obtainable using kneading compaction in the field.

A field test pad involves construction of several lifts of soil compacted similar to the construction design for the compacted soil barrier. This procedure allows the CQC personnel to verify that the laboratory-developed moisture–density–permeability relationship can be reproduced in the field. It also allows CQC personnel to check the ability of proposed construction equipment to bond lifts together, to break down clods of soil and blend them into a continuous mass and to deliver the compactive effort necessary to obtain the density required for an acceptable permeability and to determine the number of passes with the compaction equipment required to obtain the required density.

Day and Daniel (1985) identified a significant difference between the permeability measured on laboratory samples collected from a compacted soil barrier and that measured directly using a large scale field test because laboratory samples are small and are unlikely to contain some of the hydraulic defects that would be measured in a large scale test. As mentioned previously, comparable soil densities obtained using different compactive methods and efforts which could occur in the laboratory and in the field, will have different soil fabrics with different permeabilities (Mitchell, 1993).

Goldman et al. (1990) suggest performing a field permeability test on the field test pad to ensure that the compactive effort applied in the field will result in a permeability comparable to that measured in the laboratory. These tests are lengthy, sometimes up to several weeks (Daniel, 1989), and therefore are not practical for routine monitoring during construction of the barrier layer. This test involves construction of a field infiltrometer on the test pad and collection of data from this device over some period of time.

b. Construction. During construction, the CQC team observes the construction process, collects soil samples for laboratory testing, and performs field tests according to the CQC plan. These procedures are designed to determine if the soil has been properly placed and to identify potential defects in the cover barrier. The subgrade soils that serve as the foundation for the barrier caps are prepared before construction of the barrier. Depending on the type of waste being contained, a gas collection system might be constructed. The degree to which a cap limits infiltration depends on the slope of the constructed cover surface. If the waste is sufficiently deep, grades are established by placement of the subgrade soils and the final cap is constructed in uniformly thick layers over the graded subgrade. If the waste deposit is shallow, imported fill may be required to achieve the desired surface grades. Optical survey methods are generally used to measure conformance with specified grading plans.

Loose lifts of barrier soil are placed over the prepared subgrade. The thickness must be controlled to that identified in the field test pad construction. Often grade stakes, marked at the specified loose lift thickness, are placed in a grid across the portion of the cap being constructed to assist equipment operators spreading the soil. Optical survey measurements are made following compaction to document that the lift thickness has complied with specifications. Care must be taken with this method to fill the holes left following removal of the grade stakes. This can be done with bentonite powder or by rolling the area until the hole has collapsed.

After each soil lift has been compacted, specified locations are tested for moisture content and dry density. These tests can be done using either a nuclear density gauge or a sand cone device. However, considering the frequency (Goldman et al., 1990) required for these tests, a sand cone device is considered impractical due to the length of time required to perform the test. Routinely, samples of barrier soil are collected from next to the in-place field density measurements, and these samples are oven dried to measure the moisture content for comparison with and calibration of the nuclear density gauge moisture content measurements (USEPA, 1991a).

These field test data are compared to the laboratory generated Acceptance Zone, which may be modified considering the field test pad observations and data. Field tests that do not fall within the Acceptance Zone are rejected and the corrective measures specified in the CQC plan are implemented. The usual procedure is to reject areas associated with data that do not comply with the Acceptance Zone (USEPA, 1991a).

During compaction the CQC personnel observe the soil to see that clods are properly destroyed and that desiccation cracks are not developing. Either of these conditions are cause for rejection, with subsequent corrective action specified in the CQC plan. Visual observations are often the most important form of CQC (Goldman et al., 1990).

In addition to making field measurements, CQC personnel collect undisturbed samples of the compacted barrier soil. These samples consist of Shelby tube or block samples and are used to measure the permeability, dry density, and moisture content. The frequency of sampling is recommended by the USEPA (1991a). Holes made by this sampling should be filled with barrier soil that is tamped or with bentonite grout.

The final thickness of the barrier is measured using optical survey techniques. However, if the foundation soil is compressible, the measurements may significantly underestimate the actual barrier thickness and another technique, such as probing with a small diameter auger may be appropriate. These auger holes should also be filled.

Goldman et al. (1990) recommend that following construction, a field permeability test be done using a sealed double-ring infiltrometer. In this test a larger portion of the barrier is evaluated than can be represented by Shelby tube or block samples.

The completed barrier must be protected from weathering. Goldman et al. (1990) state that the CQC personnel should inspect the compacted barrier surface for damage from desiccation cracking before it is covered with either a subsequent lift of barrier soil or with a protective soil layer. It should also be checked for freezing. These weathering mechanisms increase soil permeability. Barriers affected by these occurrences should be repaired by scarifying and recompacting the affected area.

5.6.2 Compacted Soil Augmented with Bentonite

Similar CQC procedures are used for barriers constructed with bentonite augmented soil as for those constructed with nonaugmented soil. This section identifies the differences in the CQC techniques used for the two materials.

a. Preconstruction. During preliminary subsurface explorations of barrier soil borrow sources, it may become apparent through visual observations or through laboratory analyses that the required permeability cannot be achieved. In this case, augmentation of the borrow material with bentonite may be a viable option, depending on the economics of the particular project.

The USEPA (1991a) describes the following procedure for determination of the amount of bentonite required for bentonite-soil mixes. Varying amounts by weight of bentonite are added to the proposed soil and the moisture–density relationship is measured for each mix. Samples of each mix are prepared that are 2% wet of the optimum moisture content, and these samples are tested for permeability to develop a bentonite content–permeability relationship. The data are reviewed and a bentonite content is selected (with a factor of safety added to account for field variability), and the procedure described previously is followed to determine the Acceptance Zone.

The properties of different bentonite sources vary and, therefore, the properties of the design mixes should be known so that they can be specified as minimum requirements. Since bentonite is imported to the site, it should have attached certificates of compliance (Goldman et al., 1990). Additionally, the bentonite properties should be measured in the field using test methods described by the API (1990).

b. Construction. The CQC procedures for augmented soil caps are nearly identical to those used for soil caps. Field tests are made to measure the in-place moisture content and dry density and these data are compared to the Acceptance Zone. Undisturbed samples are collected and laboratory tested for moisture content, density, and permeability.

Augmented soil caps generally are susceptible to desiccation cracking, particularly during summer months. This cracking can be limited by covering the completed lift with a subsequent lift or with a protective layer soon after it is determined to be acceptable. These caps are also susceptible to the other weathering mechanisms described previously and affected zones should be remediated before covering.

5.6.3 Geomembranes

As discussed previously, geomembranes are used together with compacted soil to form a composite cap which complies with the requirements of RCRA regulations (USEPA, 1989c), or may be used alone when there is no economically suitable soil available. As the purpose of a cap is to retard infiltration of precipitation, compatibility with various chemicals is not a design or CQC consideration (USEPA, 1989c). Rather, the mechanical properties dictate the choice of geomembrane material. This section is generic to geomembranes rather than to a specific type of geomembrane material. The discussion describes the testing that is done before construction which involves testing material properties as well as that done during construction which involves testing the seams and inspecting the geomembrane for defects.

a. Preconstruction. One of the first activities undertaken for CQC of a geomembrane cover is the preparation of a CQC plan, which describes the parties involved in the manufacture and construction of the geomembrane. The various roles and requirements (in terms of providing quality to the constructed geomembrane) of the manufacturer, the installer, and those monitoring (testing) the geomembrane are described elsewhere (Koerner and Daniel, 1992). The plan outlines testing that is done by the manufacturer (referred to as manufacturer quality control), that which is done by the owner's representative (the manufacturing quality assurance personnel), that which is done by the installer (called construction quality control), and that which is done during installation by the owner's representative (construction quality assurance).

The plan specifies the personnel requirements of the manufacturer, the installer, and those monitoring manufacture and installation. Testing methods and frequencies during manufacturing and installation are specified for both parties. Documentation requirements are also described. In addition, the CQC plan specifies the corrective action(s) to be taken for seams that do not meet specifications. This usually calls for replacing the defective seam and retesting it or patching the seam and testing the patch.

The manufacturer is responsible for testing the geomembrane for mechanical properties such as thickness, tensile strength, puncture, and tear resistance. The material is examined for the absence of holes, and other site and material specific tests are done. Seams made in the factory are tested by the manufacturer. The USEPA (1986) states that the entire length of each seam should be nondestructively tested.

The CQC plan may require that the manufacture of the geomembrane be observed and tested (or that tests done by the manufacturer be monitored) by the quality assurance (QA) representative. All rolls of geomembrane are assigned an identification number, printed on a tag, and attached to the geomembrane roll. The CQC personnel inspect each roll delivered to the site for the identification tag and review the test data for that roll.

b. Construction. Each roll of geomembrane is inspected when it arrives on the construction site. If chemical compatibility is a concern, a sample of each roll should be collected and "fingerprint" tested. The rolls should be observed for damage that may have occurred in shipping and the suitability of the storage area for the geomembrane should be ascertained.

The surface of the subgrade for the geomembrane should be inspected for sharp objects, and since the subgrade is often the compacted soil barrier, the surface should be inspected for desiccation cracking. Any cracking should be repaired before placing the geomembrane.

The roll identification tags are checked as rolls of geomembrane are deployed to verify that the roll satisfied all of the manufacturer's quality control requirements. The requirements of the CQC plan are followed. The plan may contain restrictions on ambient weather conditions during placement, placement around appurtenances, such as gas vents or anchor trenches, and special procedures for calibrating seaming equipment. The geomembrane is carefully observed once it has been deployed, visually inspecting it for defects such as nicks, gouges, holes, and tears. All defects are located and marked so that they can be located later and repaired.

Before beginning field seaming, trial seam strips are made (USEPA, 1991b) to assess the equipment that is being proposed, the seaming operator, and the proposed procedures. Trial seams are made on pieces of excess material to simulate actual seaming conditions. This technique allows the seaming procedure to be evaluated before production seaming begins on the geomembrane sheet, thus avoiding potential damage to the sheet. Trial seams are made each morning, after extended breaks (e.g., lunch) and after each change in seaming personnel. Samples are taken from each test strip and mechanically tested to evaluate the seam. The samples are destructively tested for shear and peel in the field when the installer believes that sufficient strength has developed in the seam.

Production field seams should be observed by CQC personnel, noting any defects that occur. The entire length of each seam should be nondestructively tested (USEPA, 1986). Field seam types depend on the parent material type. One nondestructive test utilizes a vacuum box (ASTM D4437) and can be applied to the testing of most seam types. Pressure and other tests are also used and may be generic to a certain seam type (Koerner, 1990).

Construction quality control programs also specify collecting and destructively testing samples cut from a constructed seam. The spacing of these samples varies, depending on the project requirements.

5.6.4 Geosynthetic Clay

Geosynthetic clay liners (GCLs) consist of geosynthetic materials (geomembrane or geotextile) and bentonite as described previously. The CQC for GCLs is similar to that of geomembranes in that quality control begins at the factory where the GCL is produced. This section presents the CQC methods utilized at the place of manufacture (preconstruction) as well as the procedures used in the field during construction.

a. Preconstruction. Preconstruction CQC consists of sampling and testing the materials used to manufacture the GCL. Various manufacturers have described their respective procedures (Heerten et al., 1992; Grube, 1992b; Bruton,

1991). Each commercial product employs somewhat different materials, and therefore the specific CQC approach is dependent on the actual materials used. Two products (Bentofix® and Claymax®), where published data are available, are discussed below.

Bentofix®. The manufacturer of Bentofix® tests the bentonite used in the product, the geotextile and the final product (Heerten et al., 1992) as part of the manufacturing quality control program. The bentonite is tested for swell properties (using the United States Pharmacological NFXVII method), moisture absorption (ASTM E946, although a preferred method is the Enslin-Neff test), and moisture content (ASTM D4643). The swelling properties are related to the ability of the bentonite to seal voids. The moisture absorption is a measure of the bentonite quality. The greater the moisture content, the less swell potential for a given bentonite. The manufacturer also measures the amount of montmorillonite clay in the bentonite using either X-ray diffraction or the methylene blue test.

Geotextiles are tested using standardized tests including mass per unit area (ASTM D3376), thickness (ASTM D1777), grab tensile strength (ASTM D4632), trapezoidal tear strength (ASTM D4533), and burst strength (ASTM D3786).

The final GCL product is tested for mass per unit area, thickness, grab tensile strength, puncture resistance, peel strength, moisture content, and permeability. Heerten et al. (1992) recommend test methods, testing frequency, and minimum value levels.

Claymax®. Claymax® differs from Bentofix® in that it has adhesive edges for seaming. The manufacturer tests the bentonite, geotextile, adhesive, and final product. The bentonite supplier provides independent tests and a certificate of compliance with delivery of the bentonite material. The Claymax® manufacturer also tests the bentonite for gradation, moisture content, pH, plate water absorption, free swell, and fluid loss. Approximately 8–10 tests are performed per rail car. Additionally, the raw material is subjected to X-ray diffraction (however, the frequency of testing is not stated by Grube, 1992b). A sample of bentonite is removed from the finished product and tested for the above properties (except gradation and X-ray diffraction) for every 20,000-ft^2 (test methods are ASTM or slightly modified API or USP-NF).

Geotextiles have certificates of compliance submitted by the supplier, attesting that the material has specified values for grab tensile strength, grab tensile elongation, puncture strength, Mullen burst, mass per unit area, and wide width tensile. The geotextile is tested for these properties upon receipt from the supplier for every 7–10 delivered truck loads. The test methods are described in ASTM D4354 and D4759.

The adhesive material used to bond the geotextile and the bentonite is also tested. Raw materials used to make the adhesive are received with certificates from the supplier, documenting that the materials meet specifications. Additionally, the final adhesive is sampled and tested for pH, viscosity, and other properties.

The final composite product is sampled every 20,000 ft^2 (or as prescribed by ASTM D4354), and the bentonite from this sample is tested for bentonite content, moisture content, free swell, and fluid loss. The sample is tested for permeability (ASTM D5084). Additionally, GCL sheets are overlapped in a 30-cm diameter cell and tested for permeability. The sheets are also tested for GCL swell and the plate water absorption of removed clay.

b. Construction. Procedures that are implemented for CQC during construction are not described in the published literature. Bruton (1991) implies that field CQC is limited to observing seams to see that they satisfy the specified overlap in addition to simple visual inspection for holes or other damage. This is all that is required because of the swelling nature of the bentonite.

5.6.5 Postclosure Monitoring

Postclosure monitoring techniques for CERCLA covers are described by the USEPA (1991a). Monitoring the quantity and quality of leachate from a closed facility provides an indication of the performance of the cover system. This technique is applicable to the four types of covers described. The leachate quantity generated in a facility with a properly designed and constructed cover will decrease with time. A facility with an improper cap will have leachate generation rates that reflect precipitation trends. Monitoring the leachate quality can also provide an indication of the cap effectiveness. The contaminant concentration will decrease with time to a steady-state level in a facility with a properly constructed cap. A sudden increase in the contaminant concentration indicates a break in the cap.

Gas monitoring can also provide an indication of the cap effectiveness. The rate of gas generation should decrease with time if moisture does not reach the waste. Additionally, the cap can be surveyed with a gas meter and gas detected in nonvented areas provides an indication of a breach.

6 Conclusions

This review and evaluation of containment technologies was carried out in the context of their application to the remediation of hazardous waste disposal sites and closure of RCRA solid waste management units. The successful application of these technologies to contain and control the migration of contaminants at a waste site offers a potentially cost-effective management option in situations where other alternative remediation options are presently cost-prohibitive or are near the application stage but not yet proven or available.

There are also other potentially important ancillary benefits that may be derived from successful containment applications, including the opportunity to consider other options in parallel or at an appropriate future time when more cost-effective contaminant source degradation technologies are available. The application of potentially more effective technologies still under development, or if necessary, using the contained subsurface system as an *in situ* reactor chamber, can be considered. In this regard, once containment has been accomplished, the possibility of applying *in situ* bioremediation coupled with strategically located injection and extraction wells for maintaining inward hydraulic gradients at the barrier walls while supplying the needed nutrients and coreactants can be fully explored for the site-specific conditions. Under these conditions, extracted groundwater may be recirculated to the subsurface reactor chamber for continued treatment, if necessary.

The design, construction, and monitoring of a containment project involves many professionals and skilled workers. Geotechnical engineers and hydrogeologists bring the expertise and experience needed for site characterization and safe and effective construction in the subsurface. Geotechnical engineers and hydrogeologists, working in concert with construction contractors, have enabled the successful construction of surface caps and sidewall barriers under a variety of site-specific conditions within the United States and abroad, both as seepage barriers and for controlling migration of contaminants from a source. At sites where naturally occurring clay or rock formations can serve as effective bottom barriers, comprehensive containment becomes feasible by keying sidewall barriers into the bottom formation and into a constructed cap at the ground surface.

The construction of a bottom barrier using engineered materials can be considered when naturally occurring bottom formations are unsuitable or at impractical depths below the ground surface. Although several techniques have been advanced for constructing bottom barriers, the state-of-the-art in this regard lags that for design and construction of sidewall barriers and surface caps. In view of the difficulty in fully characterizing subsurface conditions and the inherent difficulties associated with ensuring quality of construction carried out remotely from the surface, it is not surprising that construction of bottom barriers is an underdeveloped technology. However, the need for constructing bottom barriers has only recently been fully appreciated. It can be anticipated that design, construction, and monitoring methods will improve based on experience, ingenuity, and the findings from well-conceived research and development efforts.

In this regard, some specific research and development needs have been identified as a result of this review and evaluation of containment technologies and are presented here for consideration. These proposed research and development efforts have been grouped according to the type of containment technology to which they apply, that is, surface caps, sidewall barriers, and bottom barriers. Some of the proposed research and development efforts apply to more than one type of containment technology and this has been noted accordingly.

6.1 BARRIER MATERIAL FORMULATIONS

There exists the potential to improve the performance of barriers through the introduction of new or improved slurry mixes or grout formulations. The development of additives to slurry mixes and grouts or the design of completely new formulations that further reduce advective and diffusive transport, and improve the chemical resistance of the barrier formulation need to be investigated. Organophilic clays, zeolites, and high carbon fly ash may be effective additives to barrier formulations for certain contaminant species. The concept of developing an effective bioactive barrier formulation also needs to be carefully examined and researched, if found to be promising. The synthesis of new formulations based on polymer chemistry offers possibilities. New or improved formulations need to be carefully tested so that they meet the necessary site specific performance requirements including permeability, diffusivity, workability, durability, chemical resistance, retardation, and affordability.

There is also a need to develop an improved predictive model for quantifying the expected performance of barrier formulations after placement. Such a model must account for the various processes affecting the transport and fate of contaminant species that contact the barrier under anticipated field conditions. This would include alteration of the hydraulic conductivity, effective diffusivity, sorptive properties, and chemical resistance of the barrier formulation with time due to the presence of the various contaminant species. Research efforts to develop new or improved formulations including laboratory testing, and better predictive models for quantifying performance should address the most critical contaminant species in terms of prevalence and candidacy for containment application.

6.2 VERTICAL SIDEWALL BARRIERS

Construction methods for installing vertical sidewall barriers deserve careful reexamination. For example, the vibrating beam has seen little use as a primary means to construct sidewall barriers because of its relatively thin cross section and the difficulty in penetrating dense formations. However, this technique may be excellent for installing a thin organophilic clay (or zeolite) layer within a conventional soil–bentonite barrier wall.

More needs to be known about the state-of-stress in a constructed sidewall barrier. Well-designed laboratory experiments combined with field measurements in constructed barrier walls should lead to better designs in the future.

The detrimental effects of freeze–thaw cycles on clay mixtures are well documented and can be addressed during construction. However, the effects of fluctuating water tables on the long-term performance of the wall are not as well understood and there has been little done to devise measures to mitigate the potential deleterious effects that might occur. This is a topic needing investigation.

6.3 BOTTOM BARRIERS

Although there are a few technologies available for constructing bottom barriers, they have not been extensively applied for remediation. Vertical jet grouting and vertical permeation grouting have no lateral or areal limitations, but their use would involve drilling through the contaminated zone at a waste site.

Promising techniques, which are presently being considered, include the use of directional drilling combined with permeation grouting or jet grouting or combined with a technique that mixes *in situ* the parent soil with the grout. These techniques permit the construction of the bottom barrier from the surface, thus avoiding excavation into potentially contaminated soils. Modifications of these techniques are currently being investigated and suggest that placement of these systems might not be cost prohibitive. The Zublin method, a technique designed to place a continuous composite liner system beneath a waste site, although technically attractive, appears complex, uncertain, and may not be economically feasible.

In relation to the continued advancement of bottom barrier technologies, the following studies are recommended:

- Exploration of the ramifications of drilling through the contaminated zone at a waste site for the purposes of constructing a bottom barrier. Investigate the site conditions and types of contaminants through which vertical drilling would be permissible.
- Investigation of the economics of constructing a 'shallow' bottom barrier

under a waste site versus extending the vertical sidewall barriers to a relatively deep natural bottom aquiclude.

- Investigation of the distances that can be economically reached using horizontal drilling, combined with permeation grouting, jet grouting, or other techniques.
- Careful monitoring of developments in directional drilling techniques and their combination with grouting or *in situ* mixing. This is a promising technique, since it does not require drilling through contaminated zones nor does it require excavation (as are necessary for grouting galleries or trenches).
- Investigation of the necessary requirements for obtaining watertight connections between sidewall barriers and naturally occurring formations and/or constructed bottom barriers.
- Development of improved leak detection methods applicable to all barriers, including methods for repairing defects in the barriers.

6.4 SURFACE CAPS

With regard to research and development related to caps, the most pressing needs are as follows:

- More information is needed on how differential settlement influences the components of a cover system (particularly the compacted clay component) and how much differential settlement a cap can withstand. This information is needed to determine the optimum materials for a cap at a given site with a predictable settlement potential. This information can be developed only under large-scale laboratory tests or field scale tests in which differential settlement of cap components can be created and carefully monitored.
- More information on the influence of wet–dry cycles on cracking and overall performance of caps is needed. It is not known how much cover soil is needed to protect a barrier layer from desiccation. This information should be developed by a combination of laboratory and field scale tests.
- More information is needed on alternative materials for caps. Waste materials, geosynthetic clay liners, and other materials show promise for caps, but experience is limited. Ultimately, field test plots are necessary to verify performance in real-world situations.
- More information is needed on actual performance of existing caps. This information can be developed partly by collecting and analyzing data from existing cap installations, but some additional instrumentation will probably be needed.

To meet these objectives, it is recommended that a test cap be constructed. The purpose of the test cap would be to develop field data to verify that cap performance objectives can be met. The suggested size of the test cap is approximately 1 acre. The cap could be divided into four or more segments to study different configurations, materials, thicknesses, or other variables. Performance should be monitored for 5 years, although the first 2 years should provide sufficient data for preliminary conclusions to be drawn.

6.5 CONSTRUCTION QUALITY CONTROL

There are needed improvements in construction quality control and in monitoring the performance of the constructed containment barrier system. While the private sector may have various means to address these needs, the regulatory agencies and technical societies bear the primary responsibility for developing standardized procedures for quality control, testing, and monitoring performance.

Standardization of procedures for testing the compatibility of barrier wall formulations with contained chemical contaminants is needed. Similarly, standards do not exist for other activities undertaken to verify barrier wall performance, particularly sampling methods or *in situ* test methods. Verification of the integrity of an installed barrier system presents a particular challenge, especially for bottom barriers. The application of nondestructive geophysical methods needs to be further explored. Resistivity, conductivity, and ground-penetrating radar all appear to offer potential for improving the ability to assure the quality of the installed barrier system and verify its integrity.

6.6 LONG-TERM PERFORMANCE OF BARRIER SYSTEMS

Methods need to be developed and verified for predicting the long-term effectiveness of barrier systems for preventing or attenuating the migration of contaminants from contaminant source areas (Inyang, 1992). Appropriate monitoring techniques and nondestructive *in situ* investigation methods for detecting and repairing flaws in the containment barrier need to be further developed and evaluated. The long-term effects of physical, chemical, and biological processes that may occur within the containment system and affect its performance need to be investigated. The interaction of chemical contaminants with barrier materials is not well understood and the potential for improving barrier material formulations should be assessed. Mathematical models that simulate the fate and transport of contaminants across barrier walls need to be developed and their predictive capability evaluated based both on laboratory experiments and reliable field data.

References

Acar, Y. B., A. Hamidon, S. Field, and L. Scott (1985), "The Effects of Organic Fluids on Hydraulic Conductivity of Compacted Kaolinite," *Hydraulic Barriers in Soil and Rock, ASTM 874*, A. I. Johnson, R. K. Frobel, N. J. Cavalli, and C. B. Petterson, eds., American Society for Testing and Materials, Philadelphia, PA, 171–187.

Acar, Y. B. and E. D'Hollosy (1987), "Assessment of Pore Fluid Effects Using Flexible Wall and Consolidation Permeameters," *Geotechnical Practice for Waste Disposal '87*, ASCE Geotechnical Special Publication No. 13, 231–245.

Alther, G. R., J. C. Evans, K. A. Witmer, and H.-Y. Fang (1985), "Influence of Inorganic Permeants Upon the Permeability of Bentonite," *Hydraulic Barriers in Soil and Rock, ASTM STP 874*, A. I. Johnson, R. K. Frobel, N. J. Cavalli, and C. B. Petterson, eds., American Society for Testing and Materials, Philadelphia, PA, 64–74.

American Petroleum Institute (1990), "Recommended Practice for Standard Procedures for Field Testing Waster-Based Drilling Fluids," *API Recommended Practice 13-B-1*, Dallas, TX.

Anderson, D. C., W. Crawley, and J. D. Zabick (1985), "Effects of Various Liquids on Clay Soil: Bentonite Slurry Mixes," *Hydraulic Barriers in Soil and Rock, ASTM STP 874*, A. I. Johnson, R. K. Frobel, N. J. Cavalli, and C. B. Petterson, eds., American Society for Testing and Materials, Philadelphia, PA, 93–103.

Anderson, W. F., I. C. Pyrah, and S. J. Fryer (1991), "Clay Calibration Chamber for Testing Field Devices," *Geotech. Testing J., ASTM*, **14**(4), 440–450.

Arnould, M., C. Clement, D. Gouvenot, and R. Struillou (1991), "New Sorbing Grouts for Radioactive and Toxic Heavy Metals," *Eng. Geol.*, **30**(1), 127–139.

ASCE (1990), "Deep Soil Mixing Builds Cutoff Wall," *Civil Engineering*, May, p. 27.

Ballivy, G., J. C. Colin, and M. Mnif (1992), "Effectiveness and Injected Cement Grout Under Harsh Environmental Conditions," *Geotechnical Special Publication*, 1, (30) ASCE, New York, 588–600.

Barvenik, M. J. (1985, 1986), "Methods for Physical Containment of Groundwater Contamination and Field Applications of Active and Passive Physical Containment of Contaminated Groundwater," lectures and notes presented as part of a short course "Corrective Action for Containing and Controlling Groundwater Contamination," sponsored by the National Water Well Association.

Barvenik, M. J. (1992), "Design Options Using Vertical Barriers Systems," presented at ASCE International Convention and Exposition, Environmental Geotech Symposium, New York, September, 1992.

Bierck, B. R., and W. C. Chang (1994), "Contaminant Transport Through Soil–Bentonite Slurry Walls: Attenuation by Activated Carbon,'' paper presented in Miami, FL, March 6–9, 12 p.

Bodocsi, A. and M. T. Bowers (1991), ''Permeability of Acrylate, Urethane, and Silicate Grouted Sands with Chemicals,'' *ASCE Geotech. Eng. Div.*, **117(GT8),** 1227–1244.

Bodocsi, A., M. T. Bowers, and R. A. Scherer (1987), ''Waste Contamination Effects on Grout Barriers,'' *Geotechnical Practice for Waste Disposal '87*, ASCE Geotechnical Special Publication No. 13, 306–319.

Boschuk, J. (1991), ''Landfill Caps—An Engineering Perspective,'' *Geotech. Fabrics Rep.*, **9(4),** 23–34.

Bowders, J. J., Jr. (1988), ''Discussion of 'Termination Criteria for Clay Permeability Testing,' '' *ASCE J. Geotech. Eng. Div.*, **114(8),** 947–949.

Bowders, J. J., D. E. Daniel, G. P. Broderick, and H. M. Liljestrand (1986), ''Methods for Testing the Compatibility of Clay Liners with Landfill Leachate,'' *Hazardous and Industrial Solid Waste Testing: Fourth Symposium, ASTM STP 886*, American Society for Testing and Materials, Philadelphia, PA, 233–250.

Bowers, M. T., A. Bodocsi, and D. A. Carson (1988), ''Reactivity of various grouts to hazardous wastes and leachates—Phase IV,'' *Final Report*, Contract No. 68-03-3210-13 and 68-03-3379-06, U.S. Environmental Protection Agency, Cincinnati, OH.

Boyes, R. G. H. (1975), *Structural and Cutoff Diaphragm Walls*, Applied Science Publishers, Ltd., London.

Brooker, E. W. and H. O. Ireland (1965), ''Earth Pressures at Rest Related to Stress History,'' *Can. Geotech. J.*, **II(1),** 1–15.

Brown, K. W. and D. C. Anderson (1983), ''Effects of Organic Solvents on the Permeability of Clay Soils,'' Environmental Protection Agency, EPA-600/2-83-016.

Brunsing, T. P. (1983), ''Demonstration of the Block Displacement Method at Whitehouse, Florida,'' *Land Disposal of Hazardous Waste, Proceedings of the Ninth Annual Research Symposium*, U.S. Environmental Protection Agency, Cincinnati, OH.

Bruton, D. (1991), ''Bentonite Mats Meet Secondary Containment Standards,'' *Geotech. Fabrics Rep.*, **9(4),** 26–27.

Burke, G. K. (1993), personal communication.

Burke, G. K. and G. T. Brill (1993), ''Anchored Cutoff Structure Design and Construction,'' *Proceedings of the International Conference on Case Histories in Geotechnical Engineering*, ASCE, St. Louis, MO.

Cadena, F. (1989), ''Use of Tailored Bentonite for Selective Removal of Organic Pollutants,'' *J. Environ. Eng.*, **115(4),** 756–767.

Case International Company (1982), ''Case Slurry Wall Notebook,'' Manufacturers Data, Case International Company, Chicago, IL.

Cline, J. F., Da. A. Cataldo, W. E. Skiens, and F. G. Burton (1982), ''Biobarriers Used in Shallow Burial Ground Stabilization,'' *Nucl. Technol.*, **58(2),** 150–153.

Cooley, B. (1991), ''Behavior of Soil-Bentonite Slurry Trench Cutoff Walls,'' Honor's Thesis, Bucknell University, Lewisburg, PA, December, 1991.

Corser, P. and M. Cranston (1991), ''Observations on Long-Term Performance of

Composite Clay Liners and Caps,'' *Proceedings, Geosynthetics Design and Performance*, Vancouver Geotechnical Society, Vancouver, BC.

Daniel, D. E. (1987), ''Earthen Liners for Land Disposal Facilities,'' *Proceedings, Geotechnical Practice for Waste Disposal '87*, R. D. Woods, ed., Ann Arbor, MI, June, 1987, American Society of Civil Engineers, New York, 21–39.

Daniel, D. E. (1989), ''In-Situ Hydraulic Conductivity Tests for Compacted Clay,'' *ASCE J. Geotech. Eng.*, **115(9),** 1205–1226.

Daniel, D. E. (1990), ''Summary Review of Construction Quality Control for Earthen Liners,'' *Geotechnical Special Publication No. 26*, R. Bonaparte, ed., ASCE, 175–189.

Daniel, D. E. (1993a), ''Clay Liners,'' *Geotechnical Practice for Waste Disposal*, D. E. Daniel, ed., Chapman & Hall, London.

Daniel, D. E. (1993b), personal communication.

Daniel, D. E., D. C. Anderson, and S. S. Boynton (1985), ''Fixed-Wall versus Flexible Wall Permeameters,'' *Hydraulic Barriers in Soil and Rock, ASTM STP 874*, A. I. Johnson, R. K. Frobel, N. J. Cavalli, and C. B. Petterson, eds., American Society for Testing and Materials, Philadelphia, PA, 107–126.

Daniel, D. E. and C. H. Benson (1990), ''Water Content—Density Criteria for Compacted Soil Liners,'' *ASCE J. Geotech. Eng.*, **16(12),** 1821–1830.

Daniel, D. E. and P. M. Estornell (1990), ''Compilation of Information on Alternative Barriers for Liner and Cover Systems,'' U.S. Environmental Protection Agency, Cincinnati, OH.

Daniel, D. E., S. J. Trautwein, S. S. Boynton, and D. E. Foreman (1984), ''Permeability Testing with Flexible Wall Permeameters,'' *Geotech. Testing J.*, **7(3),** 113–122.

D'Appolonia, D. J. (1980), ''Soil-Bentonite Slurry Trench Cutoffs,'' *ASCE J. Geotech. Eng. Div.*, **106(4),** 399–417.

Davis, P. M. (1983), ''Surface Deformation Associated With a Dipping Hydrofracture,'' *J. Geophys. Res.*, **88(B7),** 5826–5834.

Day, S. (1992), GeoCon, Inc. project data, personal communication.

Day, S. R. and D. E. Daniel (1985), ''Hydraulic Conductivity of Two Prototype Clay Liners,'' *ASCE J. Geotech. Eng.*, **111(8),** 957–970.

Delage, P. and H. F. Mazeas (1993), ''Surveying and Construction in Urban, Suburban and Polluted Areas—Surveying,'' *General Reports, The Environment and Geotechnics: From Decontamination to Protection of the Sub-Soil*, Paris, April 6–8.

Dougherty, D. K. (1993), Brown and Root Environmental, Inc., personal communication.

Duncan, J. M., R. B. Seed, K. S. Wong, and Y. Ozawa (1984), ''FEADAM84: A Computer program for Finite Element Analysis of Dams,'' *Geotechnical Engineering Report No. SU/GT/84-03*, Department of Civil Engineering, Stanford University, Stanford, CA. November, 1984.

Dunn, R. J. and J. K. Mitchell (1984), ''Fluid Conductivity Testing on Fine Grained Soils,'' *J. Geotech. Eng.*, **110(11),** 1648–1665.

Dvorkin, J., G. Mavko, and A. Nur (1992), ''Dynamics of Viscous Compressible Fluid in a Fracture,'' *Geophysics*, **57(5),** 720–726.

Einstein, H. H. and M. J. Barvenik (1975), ''Grouting Applications in Civil Engineering,'' *Vol. 1, Soils Publication No. 334*, prepared for Oak Ridge National Laboratory, Department of Civil Engineering, Massachusetts Institute of Technology, Cambridge, MA.

ENR (1990), ''Cut-Off Wall Scores a Record,'' *Engineering News Record*, December, 17–18.

Estornell, P. and D. E. Daniel (1992), ''Hydraulic Conductivity of Three Geosynthetic Clay Liners,'' *ASCE J. Geotech. Eng.*, **118(10),** 1592–1606.

Evans, J. C. (1991), ''Geotechnics of Hazardous Waste Control Systems,'' Chapter 20, *Foundation Engineering Handbook, 2nd ed.*, H. Y. Fang, ed., Van Nostrand-Reinhold Company, New York.

Evans, J. C. (1993), ''Vertical Cutoff Walls,'' *Geotechnical Practice for Waste Disposal*, D. E. Daniel, ed., Chapman & Hall, London.

Evans, J. C. (1994), ''Hydraulic Conductivity of Vertical Cut-off Walls,'' *Proc. Hydraulic Conductivity and Waste Contaminant Transport in Soils*, ASTM STP 1142, ed. D. E. Daniel and S. J. Trautwein, American Society for Testing and Materials, Philadelphia, PA, 79–94.

Evans, J. C. and H. Y. Fang (1983), ''Triaxial Equipment for Permeability Testing with Hazardous and Toxic Permeants,'' *ASTM Geotech. Testing J.*, **9(3),** 126–132.

Evans, J. C. and H. Y. Fang (1988), ''Triaxial and Permeability Strength Testing of Contaminated Soils,'' *Advanced Triaxial Testing of Soil and Rock*, ASTM STP 977, American Society for Testing and Materials, Philadelphia, PA, 387–404.

Evans, J. C., H. Y. Fang, and I. J. Kugelman (1985a), ''Containment of Hazardous Materials with Soil–Bentonite Slurry Walls,'' *Proceedings of the 6th National Conference on the Management of Uncontrolled Hazardous Waste Sites*, Washington, DC, 249–252.

Evans, J. C., H. Y. Fang, and I. J. Kugelman (1985b), ''Organic Fluid Effects on the Permeability of Soil–Bentonite Slurry Walls,'' *Proceedings of the National Conference on Hazardous Wastes and Environmental Emergencies*, Cincinnati, OH, 267–271.

Evans, J. C., I. J. Kugelman, and H. Y. Fang (1985c), ''Organic Fluid Effects on the Strength Deformation and Permeability of Soil–Bentonite Slurry Walls,'' *Proceedings of the 17th Mid-Atlantic Industrial Waste Conference*, Lehigh University, Bethlehem, PA, 275–291.

Evans, J. C., G. P. Lennon, and K. A. Witmer (1985d), ''Analysis of Soil–Bentonite Backfill Placement in Slurry Walls,'' *Proceedings of the 6th National Conference on the Management of Uncontrolled Hazardous Waste Sites*, Washington, DC, 357–361.

Evans, J. C. and E. N. Manuel (1985), ''Geotechnical Property Testing of Hazardous Materials and Contaminated Soils,'' *Proceedings of the 6th National Conference on the Management of Uncontrolled Hazardous Waste Sites*, Washington, DC, November, 369–373.

Evans, J. C., E. D. Stahl, and E. Droof (1987), ''Plastic Concrete Cutoff Walls,'' *Geotechnical Practice for Waste Disposal '87*, ASCE Geotechnical Special Publication No. 13, 462–472.

Gazaway, H. N. and B. H. Jasperse (1992), ''Jet Grouting in Contaminated Soils,''

Grouting, Soil Improvement and Geosynthetics, Geotechnical Special Publication, ASCE, 1 (30), 206–214.

Gerber, M. (1992), personal communication.

Gillham, R. W. and S. F. O'Hannesin (1992). "Metal-Catalyzed Abiotic Degradation of Halogenated Organic Compounds," 1992 IAH Conference *Modern Trends in Hydrogeology*, Hamilton, Ontario, May 10–13, 94–103.

Gillham, R. W., S. F. O'Hannesin, and W. S. Orth (1993), "Metal Enhanced Abiotic Degradation of Halogenated Aliphatics: Laboratory Tests and Field Trials," 1993 HazMat Central Conference, Chicago, IL, March 9–11.

Gilman, E. F., F. B. Flower, and I. D. Leone (1985), "Standardized Procedures for Planting Vegetation on Completed Sanitary Landfills," *Waste Management Res.*, **3,** 65–80.

Gipson, A. H., Jr. (1985), "Permeability Testing on Clayey Soil and Silty Sand–Bentonite Mixture Using Acid Liquor," *Hydraulic Barriers in Soil and Rock, ASTM STP 874*, A. I. Johnson, R. K. Frobel, N. J. Cavalli, and C. B. Petterson, eds., American Society for Testing and Materials, Philadelphia, PA, 140–154.

GKN Hayward-Baker (1989a), "Case History-Oxnard Slurry Wall," Company Literature.

GKN Hayward-Baker (1989b), "Case History-Mountain View Slurry Wall," Company Literature.

Goldman, L. J., L. I. Greenfield, A. S. Damle, G. L. Kingsbury, C. M. Northeim, and R. S. Trusedale (1990), *Clay Liners for Waste Management Facilities, Design, Construction and Evaluation*, Noyes Data Corporation, Park Ridge, NJ.

Grim, R. E. (1968), *Clay Mineralogy*, McGraw-Hill Book Co., New York.

Grisham, G. (1992), personal communication.

Grube, Jr., W. E. (1990), "Measuring Performance of Clay Containment Barriers," *Proceedings of the ASCE 17th Annual National Conference "Optimizing the Resources for Water Management,"* Boston, MA, 482–488.

Grube, Jr., W. E. (1992a), "Slurry Trench Cut-Off Walls for Environmental Pollution Control," *Slurry Walls: Design, Construction, and Quality Control, ASTM STP1129*, 69–77.

Grube, Jr., W. E. (1992b), "Quality Control in the Manufacture of Adhesive–Blended Geosynthetic Clay Liners," *Proceedings of the 6th GRI Seminar, MQC/MQA and CQC/CQA of Geosynthetics*, edited by R. M. Koerner and Y. J. Shsuan, Philadelphia, PA, 96–103.

Hausner, S., D. Haerle, and C. Witolla (1991), "Einkapselung von Altlasten Nachtraegliche Herstellung einer Basisabdichtung," *Sonderdruch aus Wasserwirtschaft,* **81(5),** Franckh-Kosmos Verlags-GmbH & Co., Stuttgart, 1–8.

Hayward-Baker (1992), promotional publications.

Heerten, G., K. P. von Maubeuge, M. Simpson, and C. Mills (1992), "Manufacturing Quality Control of Geosynthetic Clay Liners—A Manufacturers Perspective," *Proceedings of the 6th GRI Seminar, MQC/MQA and CQC/CQA of Geosynthetics*, edited by R. M. Koerner and Y. J. Shsuan, Philadelphia, PA, 86–95.

Houlsby, A.C. (1990), *Construction and Design of Cement Grouting: A Guide to Grouting in Rock Formation*, Wiley, New York.

Hover, W. (1993), personal communication.

Inyang, H. I. (1991), "Hazardous Waste Facilities in Seismic Zones," USEPA Draft Report.

Inyang, H. I. (1992), "Advances and Trends in Geoenvironmental Engineering," Proceedings of the U.S.-Canada Workshop on Recent Accomplishments and Future Trends in Geomechanics in the 21st Century, University of Oklahoma, Norman, OK.

Inyang, H. I. and V. B. Myers (1993), "Geotechnical Systems for Structures on Contaminated Sites," USEPA Technical Guidance Document Draft, Office of Solid Waste, Washington, DC.

Iskander, I. (1987), "Ground Freezing," *The Military Eng.*, **516,** 455–456.

Jefferis, S. A. (1981), "Bentonite–Cement Slurries for Hydraulic Cutoffs," *Proceedings of the 10th International Conference on Soil Mechanics and Foundation of Engineering*, Stockholm, June 15–19, 1981, A. A. Balkema, Rotterdam, 435–440.

Jefferis, S. A. (1992), "Contaminant-Grout Interaction," *Grouting, Soil Improvement and Geosynthetics*, Geotechnical Special Publication, ASCE, 2 (30), 1393–1402.

Jessberger, H. L., ed. (1991), *Geotechnics of Landfills and Contaminated Land, Technical Recommendations*, European Technical Committee, ETC 8, Ernst & Sohn, Berlin, Germany.

Jessberger, H. L. (1993a), personal communication.

Jessberger, H. L. ed. (1993b). *Sicherung von Altlasten*, A. A. Balkema, Rotterdam, 85–92.

Jessberger, H. L. and K. Stone (1991), "Subsidence Effects on Clay Barriers," *Geotechnique,* **41(2),** 185–194.

Johnson, D. I. and D. H. Urie (1985), "Landfill Caps: Long-Term Investments in Need of Attention," *Waste Management Res.*, **3,** 143–148.

Karlsson, H. (1993), "Horizontal Systems Technology for Shallow-Site Remediation," *JPT*, February 1993, 160–165.

Karol, R. H. (1990), *Chemical Grouting*, Marcel Dekker, New York.

Kauschinger, J. L. (1993), personal communication.

Kauschinger, J. L., E. B. Perry, and R. Hankour (1992), "Jet Grouting: State-of-the-Practice," *Grouting, Soil Improvement and Geosynthetics*, Geotechnical Special Publication, ASCE, 1 (30), 169–181.

Kipko, E. Y., Y. A. Polozov, O. Y. Lushinkova, V. A. Lagunov, and Y. I. Svirskiy (1991), "Integrated Grouting and Hydrogeology of Fractured Rocks in the USSR," draft translation by R. E. Williams, Univ. of Idaho, Moscow, ID.

Koerner, R. M. (1990), *Designing with Geosynthetics*, 2nd ed., Prentice-Hall, Englewood Cliffs, NJ.

Koerner, R. M. and D. E. Daniel (1992), "Manufacturing and Construction Quality Control and Quality Assurance of Geosynthetics," *Proceedings of the 6th GRI Seminar, MQC/MQA and CQC/CQA of Geosynthetics*, edited by R. M. Koerner and Y. J. Shsuan, Philadelphia, PA, 1–14.

Krumholz, L. (1993), Department of Chemical Engineering, Massachusetts Institute of Technology, personal communication.

LaGatta, M. D. (1992), "Hydraulic Conductivity Tests on Geosynthetic Clay Liners

Subjected to Differential Settlement," M.S. Thesis, University of Texas, Austin, TX.

LaGrega, M. D., P. L. Buckingham, and J. C. Evans (1994), *Hazardous Waste Management*, McGraw-Hill Book Co., New York.

LaPlante, C. M. and T. F. Zimmie (1992), "Freeze/Thaw Effects on the Hydraulic Conductivity of Compacted Clays," Transportation Research Board, National Research Council, *Transportation Res. Record,* **1369,** 126–129.

Lambe, T. W. and R. V. Whitman (1979), "The Permeability of Fine Grained Soils," *ASTM STP 163*, American Society for Testing and Materials, Philadelphia, PA, 55–67.

Lehman, C. T. (1993), Representative of Kajima Engineering and Consulting, Inc., at HAZMACON 93, personal communication.

Lentz, R. W., W. D. Horst, and J. O. Upport (1985), "The Permeability of Clay to Acidic and Caustic Permeants," *Hydraulic Barriers in Soil and Rock, ASTM STP 874*, A. I. Johnson, R. K. Frobel, N. J. Cavalli, and C. B. Petterson, eds., American Society for Testing and Materials, Philadelphia, PA, 127–139.

Leonards, G. A. and J. Narain (1963), "Flexibility of Clay and Cracking of Earth Dams," *ASCE J. Soil Mech. Foundations Div.*, **89 (SM2),** 47–98.

Leonards, G. A., F. Schmednecht, J. L. Chameau, and S. Diamond (1985), "Thin Slurry Cutoff Walls Installed by the Vibrated Beam Method," *Hydraulic Barriers in Soil and Rock, ASTM STP 874*, A. I. Johnson, R. K. Frobel, N. J. Cavalli, and C. B. Petterson, eds., American Society for Testing and Materials, Philadelphia, PA, 34–44.

Lutenegger, A. J. (1988), "In-Situ Test Results," Report to J. C. Evans, Bucknell University, Lewisburg, PA, February 13, 1988.

Manassero, M. and E. Pasqualini (1993), "Surveying and Construction in Urban, Suburban and Polluted Areas—Construction," *General Reports, The Environment and Geotechnics: From Decontamination to Protection of the Sub-Soil*, Paris, April 6–8.

Manuel, E. M., J. C. Evans, and R. D. Singh (1987), "Discussion of 'Hydraulic Conductivity of Two Prototype Clay Liners' by S. R. Day and D. E. Daniel," *ASCE J. Geotech. Eng. Div.*, **113 (GT7),** 804–806.

McBride, R. A. and N. Baumgartner (1992), "Simple Slurry Consolidometer Designed for the Estimation of the Consistency Limits of Soils," *J. Terramech.* **29(2),** 223–238.

McCandless, R. M. and A. Bodocsi (1987), "Investigation of Slurry Cutoff Wall Design and Construction Methods for Containing Hazardous Wastes," Environmental Protection Agency, EPA/600-S2-87/063, Cincinnati, OH.

McCandless, R. M. and A. Bodocsi (1988), "Hydraulic Characteristics of Model Soil–Bentonite Slurry Cutoff Walls," *Proceedings of the 5th National Confernce on Hazardous Wastes and Hazardous Materials*, Las Vegas, NV, April, 198–201.

McLay, D. (1987), "Installation of a Cement–Bentonite Slurry Wall Using the Vibrating Beam Method: A Case History," *Proceedings of the Nineteenth Mid-Atlantic Industrial Waste Conference*, Bucknell University, Lewisburg, PA, June, 1987, 272–282.

McMahon, D. (1993), personal communication.

McMahon, D. R., M. J. Mann, and R. C. Tulett, (1994), "Building a Barrier Wall through Boulders," *Civil Engineering,* **64(10),** 60–62.

Meegoda, N. and R. A. Rajapakse (1994), "Short-Term and Long-Term Permeabilities of Contaminated Clays," *J. Environ. Eng.,* **119(4),** 725–743.

Meseck, H. and E. Glaser (1986), "Moeglichkeiten der Nachtraeglichen Herstellung Horizontaler Dichtungssohlen bei Deponien und Altlasten," IRC, 29, Berlin.

Millet, R. A. and J.-Y. Perez (1981), "Current USA Practice: Slurry Wall Specifications," *ASCE J. Geotech. Eng. Div.,* **107 (GT8),** 1041–1056.

Millet, R. A., J.-Y. Perez, and R. R. Davidson (1992), "USA Practice Slurry Wall Specifications 10 Years Later," *Slurry Walls: Design, Construction and Quality Control, ASTM STP 1129*, D. B. Paul, R. R. Davidson, and N. J. Cavalli, eds., American Society of Testing and Materials, Philadelphia, PA.

Mitchell, J. K. (1986), "Hazardous Waste Containment," *Groundwater Eng. Geol.,* **3,** 145–157.

Mitchell, J. K. (1993), *Fundamentals of Soil Behavior*, 2nd ed., Wiley, New York.

Mitchell, J. K., D. R. Hooper, and R. G. Campanella (1965), "Permeability of Compacted Clay," *ASCE J. Soil Mechanics Foundation Div.*, 91 (SM4) 41–65.

Mitchell, J. K. and M. Jaber (1990), "Factors Controlling the Long-Term Properties of Clay Liners," *Geotechnical Special Publication No. 26*, R. Bonaparte, ed., ASCE, 84–105.

Mitchell, J. K. and F. T. Madsen (1987), "Chemical Effects on Clay Hydraulic Conductivity," *Geotechnical Practice for Waste Disposal '87, ASCE Geotechnical Special Publication No. 13*, June, 87–116.

Mitchell, J. K. and W. A. van Court (1992a), "The Role of Soil Modification in Environmental Engineering Applications," *Grouting, Soil Improvement and Geosynthetics*, Geotechnical Special Publication, ASCE, 1 (30), 110–143.

Mitchell, J. K. and W. A. van Court (1992b), "Containment Immobilization and Containment: Barriers, Walls and Covers," Geotechnical Engineering Rept. No. UCB/GT/92-09, Dept. of Civil Engineering, Univ. of California, Berkeley, CA.

Montgomery, R. J. and L. J. Parsons (1989), "The Omega Hills Final Cap Test Plot Study: Three Year Data Summary," presented at the Annual Meeting of the National Solid Waste Management Association, Washington, DC.

Mori, A. and M. Tamura (1987a), "Hydrofracturing Phenomena in Soil Grounds," *Memoirs of the School of Science and Engineering*, Waseda University, Tokyo, Japan, 51, 33–60.

Mori, A. and M. Tamura (1987b), "Hydrofracturing Pressure of Cohesive Soils," *Soils Foundations,* **27(1),** 14–22.

Mott, H. V. and W. J. Weber, Jr. (1991a), "Factors Influencing Organic Contaminant Diffusivities in Soil–Bentonite Cutoff Barriers," *Environ. Sci. Technol.* **25(10),** 1708–1715.

Mott, H. V. and W. J. Weber, Jr. (1991b), "Diffusion of Organic Contaminants through Soil–bentonite Cut-off Barriers," *Res. J. WPCF,* **63(2),** 166–176.

Mott, H. V. and W. J. Weber, Jr. (1992), "Sorption of Low Molecular Weight Organic Contaminants by Fly Ash: Consideration for the Enhancement of Cutoff Barrier Performance," *Environ. Sci. Technol.,* **26(6),** 1234–1242.

Murdoch, L. (1992a), personal interview, Center Hill Research Facility, University of Cincinnati, Cincinnati, OH.

Murdoch, L. (1992b), ''Some Recent Developments in the Delivery and Recovery: Hydraulic Fracturing and Directional Drilling,'' *Proceedings of the ETEX 1992 Second Annual Environmental Exposition and Conference*, Washington, DC, April 7–9.

Murphy, W. L. and P. A. Gilbert (1985), ''Settlement and Cap Subsidence of Hazardous Waste Landfills,'' Environmental Protection Agency, EPA/600/2-85-035, Cincinnati, OH.

Nash, K. A. (1974), ''Stability of Trenches Filled with Fluids,'' *ASCE, J. Const. Div.*, **100(4),** 533–542.

Nonveiller, E. (1989), *Grouting Theory and Practice*, Elsevier Science Publishers B.V., Amsterdam, The Netherlands.

Nordquist, J. E., G. S. Mason, and G. L. Olsen (1986), ''Clay Liner Hydraulic Conductivity: Comparison of Field and Laboratory Tests,'' *Environmental Engineering, Proceeding of the 1986 Specialty Conference*, ASCE, Cincinnati, OH, New York, 401–406.

O'Hannesin, S. F. and R. W. Gillham (1992), ''A Permeable Reaction Wall for *In-Situ* Degradation of Halogenated Organic Compounds,'' 45th Canadian Geotechnical Society Conference, Toronto, Ontario, Oct. 25–28.

Olson, R. E. and D. E. Daniel (1981), ''Measurement of the Hydraulic Conductivity of Fine-Grained Soils,'' *ASTM STP 746*, American Society for Testing and Materials, Philadelphia, PA, 18–64.

Othman, M. A. and C. H. Benson (1992), ''Effect of Freeze-Thaw on the Hydraulic Conductivity of Three Compacted Clays from Wisconsin,'' Transportation Research Board, National Research Council, *Transportation Res. Record,* **1369,** 126–129.

Persoff, P., S. Finsterle, G. J. Moridis, and K. Pruess (1994), ''Physical Barriers Formed from Gelling Liquids: Laboratory Grout Emplacement Experiments in Unsaturated Sand,'' presented at 1994 Spring Meeting of the American Geophysical Union, Baltimore, MD, May 23–27.

Peterson, S. R. and G. W. Glendon (1985), ''Interactions Between Acidic Solutions and Clay Liners: Permeability and Neutralization,'' *Hydraulic Barriers in Soil and Rock, ASTM STP 874*, A. I. Johnson, R. K. Frobel, N. J. Cavalli, and C. B. Petterson, eds., American Society for Testing and Materials, Philadelphia, PA, 140–154.

Pierce, J. J. and K. A. Witter (1986), ''Termination Criteria for Clay Permeability Testing,'' *ASCE J. Geotech. Eng. Div.*, **112(9),** 841–854.

Polozov, Ju. A. and I. V. Popov (1985), ''Grouting of Porous Aquifers During Shaft Sinking,'' *Int. J. Mine Water,* **4(1),** 25–31.

Rector, J. W. and H. F. Morrison (1993), ''Geophysical Imaging of Emplaced Barrier Continuity for Hazardous Waste Containment,'' unpublished communication.

Reina, P. (1987), ''Slurry Trench Cutoff Wall Surrounds Plant Site,'' *Engineering News Record*, October 1, 44.

Repa, E. W. (1987), ''Evaluating Asphalt Cap Effectiveness at a Superfund Site,'' *J. Environ. Eng.*, **113(3),** 649–653.

Ressi, A. and N. J. Cavalli (1985), ''Bentonite Slurry Trenches,'' *Eng. Geol.*, **21,** 333–339.

Robbat, A. (1993), Department of Chemistry, Tufts University, personal communication.

Rutovist, J., J. Noorishad, O. Stephanson, and C. F. Tsang (1992), ''Theoretical and Field Studies of Coupled Hydromechanical Behavior of Fractured Rocks. 2. Field Experiment and Modeling,'' *Int. J. Rock Mechanics Mining Sci. Geomech. Abs.*, **29(4),** 411–419.

Ryan, C. R. (1985), ''Slurry Cutoff Walls: Applications in the Control of Hazardous Waste,'' *Hydraulic Barriers in Soil and Rock, ASTM STP 874*, A. I. Johnson, R. K. Frobel, N. J. Cavalli, and C. B. Petterson, eds., American Society for Testing and Materials, Philadelphia, PA, 9-24.

Ryan, C. R. (1987), ''Vertical Barriers in Soil for Pollution Containment,'' *Geotechnical Practice for Waste Disposal '87*, ASCE Geotechnical Special Publication No. 13, 182–204.

Schulze, D., M. J. Barvenik, and J. E. Ayres (1984), ''Design of the Soil/Bentonite Backfill Mix for the First EPA Superfund Cutoff Wall,'' *Proceedings of the 4th National Symposium and Exposition on Aquifer Restoration and Groundwater Monitoring*, sponsored by the NWWA, Columbus, OH.

Shackelford, C. D. (1994), ''Waste-Soil Interactions that Alter Hydraulic Conductivity,'' *Hydraulic Conductivity and Waste Containment Transport in Soil, ASTM STP 1142*, D. E. Daniel and S. J. Trautwein, eds., American Society for Testing and Materials, Philadelphia, PA.

Shan, H. Y. and D. E. Daniel (1991), ''Results of Laboratory Tests on a Geotextile/Bentonite Liner Material,'' *Geosynthetics 91*, Industrial Fabrics Association International, St. Paul, MN, **2,** 517–535.

Spooner, P., R. S. Wetzel, C. E. Spooner, C. A. Furman, E. F. Tokarski, and G. E. Hunt (1984), *Slurry Trench Construction for Pollution Migration Control*, Environmental Protection Agency, EPA-540/2-84-001, Cincinnati, OH.

Starr, R. C. and J. A. Cherry (1990), ''*In Situ* Barriers for Groundwater Pollution Control,'' *Prevention and Treatment of Soil and Groundwater Contamination in the Petroleum Refining and Distribution Industry*, Montreal, Quebec.

Starr, R. C. and J. A. Cherry (1994), ''*In Situ* Remediation of Contaminated Ground Water: The Funnel-and-Gate System,'' *Ground Water,* **32(3),** 465–476.

Stokes, C. (1993), Department of Chemical Engineering, University of Houston, personal communication.

Sweiden, R. (1990), ''Analysis of Soil-Bentonite Slurry Trench Cutoff Wall,'' Bucknell University, Lewisburg, PA, May 15, 1990.

Swope, G. L. (1975), ''Revegetation of Landfill Sites,'' M.S. Thesis, Pennsylvania State University, University Park, PA.

Taki, O., S. Takeshima, and D. S. Yang (1990), ''Pollution Containment Using Soil–Cement Mixed Walls,'' *Chemical Contamination at Shoreline Sites*, HAZMACON 90, Anaheim, CA, April 17–19, 1990.

Tan, S.-A., K.-M. Liang, T.-S. Tan, and G.-P. Kurunarratne (1992), ''Total Stress Probe Determination of Clay Slurry Density,'' *Geotech. Testing J.*, **15(1),** 59–64.

Tschebotarioff, G. P. and A. A. DePhillippe (1953), ''The Tensile Strength of Disturbed and Recompacted Soils,'' *Proceedings of the Third International Conference on Soil Mechanics and Foundation Engineering*, Switzerland, Vol. 1, 207–210.

Trezek, G. J. (1986), ''Grout and Slurry Walls for Hazwaste Containment: The Down Side,'' *Hazardous Waste Hazardous Materials,* **3(3),** 281–292.

Ueda, T., H. Nakasaki, S. Suzuki, T. Suzuli, and M. Takahashi (1987), ''Application of Acoustic Emission Technique to Foundation Grouting of Rock Masses,'' *Takenaka Tech. Res. Rep.,* **38,** 111–125.

U.S. Environmental Protection Agency (1983), ''Standardized Procedures for Planting Vegetation on Completed Sanitary Landfills,'' EPA/600/2-83-055, Cincinnati, OH.

U.S. Environmental Protection Agency (1984), ''Slurry Trench Construction for Pollution Migration Control,'' EPA/54/2-84-001.

U.S. Environmental Protection Agency (1986), ''Technical Guidance Document: Construction Quality Assurance for Hazardous Waste Land Disposal Facilities,'' EPA/530-SW-86-031, Cincinnati, OH.

U.S. Environmental Protection Agency (1987), ''Construction Quality Control and Post-Construction Performance Verification for the Gilson Road Hazardous Waste Site Cutoff Wall,'' EPA/600/2-87/065, Cincinnati, OH.

U.S. Environmental Protection Agency (1989a), ''Transport and Fate of Contaminants in the Subsurface,'' EPA/625/4-89/019, Cincinnati, OH.

U.S. Environmental Protection Agency (1989b), ''Requirements for Hazardous Waste Landfill Design, Construction, and Closure,'' EPA/625/4-89/022, Washington, DC.

U.S. Environmental Protection Agency (1989c), ''Technical Guidance Document, Final Caps on Hazardous Waste Landfills and Surface Impoundments,'' EPA/530-SW-89-047, Washington, DC.

U.S. Environmental Protection Agency (1991a), ''Design and Construction of RCRA/CERCLA Final Caps,'' EPA/625/4-91/025, Washington, DC.

U.S. Environmental Protection Agency (1991b), ''Technical Guidance Document: Inspection Technique for the Fabrication of Geomembrane Seams,'' EPA/530/SW-91/051, Cincinnati, OH.

Van Olphen, H. (1977), *An Introduction to Clay Colloid Chemistry*, Wiley, New York.

Walker, A. (1993), personal communication.

Ward, L. M. (1984), ''Close-up on Cleanup at Lipari Waste Site,'' *Hazardous Materials & Waste Management Magazine*, May–June, 40–42.

Weaver, D., J. C. Evans, and S. E. Pancoski (1990), ''Grout Testing for Hazardous Waste Application,'' *Concrete International*, July, 45–47.

Welsh, J. P., (1992), ''Grouting Techniques for Excavation Support,'' *Excavation and Support for the Urban Infrastructure*, T. D. O'Rourke and A. G. Hobelman, eds. Geotechnical Special Publication No. 33, ASCE New York.

Welsh, J. P. (1993), personal communication.

Welsh, J. P. and G. K. Burke (1991), ''Jet Grouting—Uses for Soil Improvement,'' *Geotechnical Engineering Congress, Vol. I*, F. G. McLean, D. A. Campbell, and D. W. Harris, eds., Geotechnical Special Publication No. 27, ASCE, New York.

Witherspoon, P. A. (1993), personal communication.

Xanthakos, P. P. (1994), *Slurry Walls as Structural Systems, 2nd edition*, McGraw-Hill Book Co., New York.

Wu, J. Y. and R. P. Khera (1990), ''Properties of a Treated–Bentonite/Sand Mix in Contaminant Environment,'' ASTM STP 1095, Philadelphia, PA, 47–59.

Yang, D. S., U. Luscher, I. Kimoto, and S. Takeshima (1993), ''SMW Wall for Seepage Control in Levee Reconstruction,'' *3rd International Conference on Case Histories in Geotechnical Engineering*, Univ. of Missouri-Rolla, MO, 487–492.

Yeung, A. T. (1990), ''Electro-Kinetic Barrier to Contaminant Transport through Compacted Clay,'' Ph.D. Thesis, University of California, Berkeley, CA.

Zimmie, T. F. (1992), ''Freeze–Thaw Effects on the Permeability of Compacted Clay Liners and Covers,'' *Geotech. News,* **10(1),** 28–30.

Zimmie, T. F., J. S. Doynow, and J. T. Wardell (1981), ''Permeability Testing of Soils for Hazardous Waste Disposal Sites,'' *Proceedings of the 10th International Conference on Soil Mechanics and Foundation Engineering*, Stockholm, Sweden, Vol. 2, 403–406.

Zimmie, T. F. and C. M. LaPlante (1990), ''The Effect of Freeze/Thaw Cycles on the Permeability of Fine-Grained Soil,'' *Proceedings of the 22nd Mid-Atlantic Industrial Waste Conference*, Drexel University, Philadelphia, PA, 580–593.

Zimmie, T. F., C. M. LaPlante, and D. L. Bronson (1991), ''The Effect of Freezing and Thawing on Landfill Covers and Liners,'' *Proceedings of the Third International Symposium on Cold Regions Heat Transfer*, Fairbanks, AK, 363–371.

Glossary

Activity The plasticity index normalized with respect to the clay fraction. High-activity clays are subject to greater ranges of shrinking and swelling than low-activity clays.

Adsorption The attraction and adhesion of a layer of ions from an aqueous solution to the solid mineral surfaces with which it is in contact.

Advection The process by which solutes are transported by the motion of flowing groundwater. Generally, the principal means for contaminant transport in groundwater.

Anisotropy The condition under which one or more of the hydraulic properties of an aquifer vary according to the direction of flow.

Aquiclude, Aquitard A naturally occurring low-permeability formation that forms either the upper or lower boundary of a groundwater flow system.

Aquifer Rock or sediment in a formation, group of formations, or part of a formation that is saturated and sufficiently permeable to transmit useful quantities of water to wells and springs.

Arching The restriction of the consolidation of soil–bentonite backfill in a trench with relatively rigid sidewalls. The consolidation of the backfill is restricted by the friction at the trench–backfill interface.

Asphalt concrete Asphaltic paving material of the type used to construct roadways.

Atterberg limits The water contents at certain limiting or critical stages in soil behavior used to correlate the engineering properties and engineering behavior of fine-grained soils.

Bedrock A general term for the rock, usually solid, that underlies soil or other unconsolidated material.

Bentonite Any commercially processed colloidal clay, largely made up of the mineral sodium montmorillonite, a hydrated aluminum silicate.

Biological degradation Removal of a contaminant species by biological oxidation.

Bit The cutting tool attached to the bottom of the drill stem.

Cap An engineered barrier placed at the ground surface to control infiltration into underlying materials, to provide a physical separation of the underlying materials from the surface environment, to control release of gas to the surface, or to serve other site-specific objectives.

Capillary fringe The zone immediately above the water table, where water is drawn upward by capillary attraction.

Cement–bentonite cutoff wall Constructed by excavating a trench filled with a bentonite slurry and displacing it with a cement–bentonite slurry or by supporting a trench with cement–bentonite slurry. The material sets up in the trench and becomes a barrier wall.

Cement–bentonite slurry A combination of bentonite clay, cement, water, and sand that results in a hardened material with a consistency of stiff clay. Frequently employed for control of groundwater seepage.

Chemical grout Grout composed of a chemical base, catalyst, and water or solvent (e.g., sodium silicate, acrylate, and urethane).

Clay Fine-grained soil type that can be made to exhibit plasticity within a range of water content and that exhibits considerable strength when dried.

Cobble A rock fragment, usually rounded or semirounded, with an average dimension between 3 and 12 in.

Colloid Extremely small solid particles, 0.0001–1 μm in size, which will not settle out of a solution; intermediate between a true dissolved particle and a suspended solid that will settle out of solution.

Compacted clay Clay that has been compacted by mechanical means, such as with specialized earth-compaction equipment.

Composite cutoff wall A barrier wall constructed using a combination of materials, such as a geomembrane liner placed within a slurry trench cutoff wall to form a barrier.

Composite liner A liner consisting of two different barrier materials in contact with each other (commonly a geomembrane in contact with clay).

Containment barrier system Passive barriers to prevent or limit contaminant migration from both contaminated industrial sites as well as from newly constructed waste disposal facilities.

Contaminant Any constituent, impurity, or pollutant that adversely affects the quality of surface water, groundwater, or soil, and/or poses a threat to human health or the environment.

CQA Construction quality assurance; usually conducted by or for the project engineer providing additional assurance to the owner that the quality of construction conforms to plans and specifications.

CQC Construction quality control; the contractor's observations, sampling, measuring, and testing to establish conformances to plans and specifications.

Deep soil mixing Method that utilizes a special auger mixing shaft that rotates into the ground while simultaneously permitting the injection of bentonite and water, or cement, bentonite, and water slurry.

Dewatering The process of lowering the groundwater level to a predetermined depth.

Diffusion A molecular process by which both ionic and molecular species dissolved in water move from areas of higher concentration to areas of lower concentration.

Dispersion The phenomenon by which a solute in flowing groundwater is mixed with uncontaminated water and becomes reduced in concentration. Dispersion is caused by both differences in the velocity that the water travels at the pore level and differences in the rate at which water travels through different strata in the flow path.

DNAPLs Dense nonaqueous phase liquids (heavier than water); typically have low solubilities in water and high interfacial surface tensions.

Effluent A waste liquid from a manufacturing or treatment process, in its natural state or partially or completely treated, that is discharged to the environment.

Evapotranspiration Movement of water from soils to the atmosphere as a result of evaporation from soil and transpiration of water from plants.

Filter A layer of material that separates two zones of differing grain sizes or prevents fine materials from migrating from one zone into another.

Gap graded Soil with an excess or deficiency of certain sizes or one with particles mostly of the same size.

Gel A state of a colloidal suspension in which shearing stresses below a certain finite value fail to produce permanent deformation. Gels commonly occur when the dispersed colloidal particles have a great affinity for the base fluid.

Geomembrane An essentially impermeable geosynthetic composed of one or more synthetic (polymeric) sheets.

Geonet A geosynthetic consisting of integrally connected parallel sets of ribs overlying similar sets at various angles for planar drainage of liquids and gases.

Geosynthetic clay liner Factory manufactured, hydraulic barrier typically consisting of bentonite clay or other very low permeability material, supported by geotextiles and/or geomembranes that are held together by needling, stitching and/or chemical adhesives.

Geosynthetic erosion control material A manufactured material spread on the surface and intended to minimize erosion of subsoils.

Geosynthetics The generic term for all synthetic materials used in geotechnical engineering applications.

Geotextile A permeable geosynthetic comprised solely of textiles.

Glacial till A general term for unconsolidated sediment transported by glaciers and deposited directly on land or in the sea.

Graded An engineering term pertaining to a soil or an unconsolidated sediment consisting of particles having a distribution of particle sizes ranging from coarse to fine.

Groundwater The water contained in interconnected pores located below the water table in an unconfined aquifer or located in a confined aquifer.

Grout A fluid mixture of cement and water of a consistency that can be forced through a pipe and placed as required. Various additives, such as sand, bentonite, and hydrated lime, may be included in the mixture.

Hanging wall A vertical barrier that does not fully penetrate an aquifer.

Head Energy contained in a water mass, produced by elevation, pressure, or velocity.

Heterogeneous Pertaining to a substance having different characteristics in different locations.

Hydraulic conductivity The rate of discharge of water under laminar flow conditions through a unit cross-sectional area of a porous medium under a unit hydraulic gradient and standard temperature conditions.

Hydraulic gradient The change in total head over a measured distance in a given direction.

Hydrofracturing The injection of fluids under pressure to form and open fractures by overcoming overburden pressure, cohesive force, and pore fluid pressure.

Hydrogeological conditions The interrelationship of geologic materials with groundwater processes unique to a given site.

Igneous rocks Rocks that solidified from molten material, that is, from a magma.

Infiltration Downward percolation of precipitated water into underlying subsoils.

Isotropic Medium whose properties are the same in all directions.

Jet grouting Pumping grout under very high pressure from the bottom of rotating rods inserted into the ground. The grout is mixed with the surrounding soil.

Key A groove channeled into a subsurface formation that allows a vertical barrier to conform and adhere to the formation.

Landfill A general term indicating a disposal site for refuse and other waste.

Leachate Water that contains contaminants created by seeping through a contaminant source area.

Leachate collection system A system installed to capture leachate.

Lift Term applied to construction of a discrete layer of a soil liner, usually 6–9 in. in thickness.

Limestone A sedimentary rock consisting chiefly of calcium carbonate, primarily in the form of the mineral calcite.

Liner A low-permeability material, such as clay or plastic sheeting, that is put beneath a landfill in order to capture any leachate generated so as to help prevent groundwater contamination or in a cap to prevent infiltration.

LNAPL Light nonaqueous phase liquid (less dense than water).

Marsh funnel A device used to measure the viscosity of a fluid.

NAPL Nonaqueous phase liquid.

OCR Overconsolidation ratio; the ratio of the preconsolidation stress to the existing vertical effective overburden stress.

Overconsolidation A soil whose preconsolidation pressure is greater than the existing overburden pressure.

Packer test An aquifer test performed in an open borehole; the segment of the borehole to be tested is sealed off from the rest of the borehole by inflating seals, called packers, both above and below the segment.

Particulate grout Grout composed of Portland Cement and/or clay and water.

Passive containment barrier A nonactive, nonoperating barrier that prevents the migration of subsurface contamination (vertical barriers, bottom barriers, and caps).

Permeability The property or capacity of a porous rock, sediment, or soil for transmitting a fluid.

Permeation grouting The construction of a barrier by pumping grout into the ground to fill voids within the geologic matrix.

Piezometer A nonpumping well, generally of small diameter, that is used to measure the elevation of the water table or potentiometric surface. A piezometer generally has a short well screen through which water can enter.

Plastic concrete A mixture of cement, bentonite, water, and aggregate.

Poisson's ratio The ratio of horizontal strain to vertical strain.

Pore space The volume between mineral grains in a porous medium.

Porosity The ratio of the volume of void spaces in a rock or sediment to the total volume of the rock or sediment.

Portland cement Hydraulic cement composed primarily of hydraulic calcium silicates that set and harden by reacting chemically with water.

Remediation The application of a process to counteract the effect of a contaminant.

ROD Record of decision.

Sand Earthen material that will pass the No. 4 (4.74 mm) sieve and be retained on the No. 200 (0.075 mm) sieve.

Sandstone A sedimentary rock composed of abundant rounded or angular fragments of sand set in a fine-grained matrix and more or less firmly united by a cementing material.

Saturated zone The zone in which the voids in the rock or soil are completely filled with water.

Shear stress That component of stress that acts tangential to a plane through any given point in a body.

Shelby tube A sampling device that is pushed into an unconsolidated formation ahead of the drill bit. Typically, the Shelby tube is pushed by hydraulic means.

Silt Earthen material that will pass the No. 200 (0.075 mm) sieve, is non-plastic or very slightly plastic, and has little or no strength when dry.

Slump test Measure of the consistency of a material, usually concrete, indicating the workability of that material.

Slurry A thin mixture of liquid, especially water, and any of several finely divided substances, such as cement or clay particles.

Slurry wall An underground wall designed to prevent or restrict groundwater flow; constructed by digging a trench and backfilling it with a slurry.

Stabilization Immobilizing a groundwater contaminant to render its impact to the environment to acceptable risk levels.

Subsidence Settlement caused by compression of underlying material.

Syneresis The exuding of water from grout that causes shrinkage of the grout and potential opening of voids.

Tortuosity The actual length of a groundwater-flow path, which is sinuous in form, divided by the straight-line distance between the ends of the flow path.

Tremie method of concrete placement A concrete placement method utilizing a long smooth, straight pipe connected to a hopper with its lower end buried in fresh concrete to maintain a seal and force the concrete into position by pressure.

Triaxial test A cylindrical compression test used to overcome some of the serious disadvantages of the direct shear testing of soil.

Volatile compounds Compounds with sufficiently low-boiling points and high-vapor pressures which, in the vapor state, are very mobile.

Water content The ratio of the amount of water present in a soil volume to the amount of soil grains based on the dry mass of the soil.

Water table The surface in an unconfined aquifer or confining bed at which the pore water pressure is atmospheric.

Well-graded Soil that has a good representation of particle sizes over a wide range.

Window A pocket of highly pervious soil material entrapped by advancing soil–bentonite backfill during the construction of a cutoff wall, resulting in a potential hydraulic defect in the wall.

INDEX

Abiotic degradation, 7
Acceptance Zone measurements, compacted soil quality control:
  bentonite augmentation, 134
  procedures for, 131–133
Acoustic emission technique:
  bottom barrier construction, jet grouting, 66–67
  permeation grouting quality control, 28
Advective transport, mechanism of, 6–7
Arching phenomenon, slurry wall state-of-stress and, 36
Asphalt paving:
  cap barrier layer with, 107, 109–110
  cap surface layer with, 84, 89-91
ASTM standards, surface caps:
  compacted soil quality control, 130–132
  construction quality control (CQC):
    Bentofix product, 137
    Claymax product, 137–138
  geomembrane composites, 136
Atterberg Limits, compacted soil quality control, 130–132

Backfill:
  slurry wall construction:
    quality control, 53–54
    wall design and, 16–17
  soil-bentonite, characteristics, 48–51
Barrier layer for final caps:
  distortion definition, 106–107
  materials for, 95, 100–110
    asphalt paving, 107, 109–110
    compacted clay, 95, 100–101, 106–107
    composite geomembrane/compacted clay, 100, 104
    composite geomembrane/geosynthetic clay, 100, 105
    geomembranes, 95, 100, 102
    geosynthetic clay, 95, 100, 103
    testing of, 131–132
    waste materials, 107–108
Barrier systems:
  active *vs.* passive, 3-4, 11
  containment technology, 1-3
  continuity, quality control and, 5
  long-term performance evaluation, 143
  material formulations, 140
  overview of applications, 3-4
  quality control, 4-5
  See also Bottom barriers; Vertical barriers
Basic soils, *See* Soils
Beam piping, Soilsaw barrier wall, 31–32
BENTOFIX liners:
  construction quality control (CQC), 137
  differential settlement measurements, 124–125
BENTOMAT liners:
  differential settlement measurements, 124–125
  shear strength testing, 122–124
Bentonite:
  backfill quality control, 54
  cement-bentonite slurry wall design and construction, 18–19
  deep soil mixing with, 23
  geosynthetic clay liners (GCLs):
    composites with, 110, 116–117
    shear strength and, 124
  hydraulic conductivity compatibility testing, 44-45
  plastic concrete slurry wall design and construction, 19–20
  slurry wall construction:
    performance and, 33–34
    quality control (CQC), 52–53
  soil-bentonite slurry, wall design and construction, 16–18
  surface control barriers (caps), compacted soil augmented with, 134
Bioactive barrier materials:
  burrowing animals and, 128
  cobbles in final cap protection layer, 91, 94
  development of, 140
Biotic degradation, 7
Block displacement method (BDM), bottom barrier construction, 70–72

Borrow soil:
compacted soil quality control, 130–131
soil-bentonite quality control, 51–52
Bottom barriers:
construction quality control (CQC), 80–82
economic factors, 78–80
future construction techniques, 141–142
grouted barriers, 61–74
block displacement method (BDM), 70–72
directional drilling and grouting, 67–70
grouts and grouting procedures, 61–63
hydrofracturing and grouting, 70
jet grouting, 65–67
permeation grouting, 63–65
sheet pile cutoff/injection grouting, 72–74
microtunneling and injection grouting, 76–77
natural barriers, 60–61
overview, 59
rectangular microtubes, 77
Zublin system, 74–76
Buffering capacity, hydraulic conductivity testing, 44–45
Burrowing animals, assessment of overall cap containment, 128

Cement. *See* Portland Cement
Cement-bentonite composite:
grouts, jet grouting, 30
slurry, composite cutoff wall construction, 23–24
construction quality control (CQC) testing, 54–56
trench cutoff wall design and construction, 18–19
vibrating beam cutoff walls, 21–22
Chemical grout, vertical barrier construction, 25
Chemicals, grouted bottom barriers, impact on, 62–63
Circumferential barrier, vertical barrier design, 11–12
Clay materials:
bottom barrier construction costs, 79–80
hydraulic conductivity compatibility and, 42–48
natural bottom barriers with, 60–61
soil-bentonite backfill, 49
CLAYMAX liners:
construction quality control (CQC), 137–138
differential settlement measurements, 124–125
shear strength testing, 122–124
Cobbles, in final surface cap design:
protection layer with, 91, 94
surface layer with, 84, 88, 90
Compacted clay liner (CCL), final cap barrier layer with, 95, 100–101, 106–107
Compacted soil:
liners (CSL), hydraulic conductivity testing, 120
surface control barriers (caps):
bentonite augmentation, 134
construction quality control (CQC), 132–133
geomembrane composites, 134–136
preconstruction quality control, 130–132
Compatibility studies:
grouted bottom barriers, grouts and grouting, 62–63
hydraulic conductivity testing, 42–48
permeation grouting, 56
slurry wall construction, 52
soil-bentonite slurry walls, 18
Composite cutoff walls, design and construction, 23–24
Conductivity. *See* Hydraulic conductivity
Consolidation pressure, slurry wall performance and, 35
Consolidometer-type permeameter, natural bottom barriers, 60–61
Construction quality assurance (CQA), containment technology, 4–5
Construction quality control (CQC):
bottom barrier construction, 80–82
containment technology, 4–5
future research in, 143
soil-bentonite cutoff walls, 51–54
backfill testing, 53–54
preconstruction phase, 51–52
slurry testing, 52–53
trench testing, 53
surface control barriers (caps), 130–138
bentonite-augmented compacted soil, 134
compacted soil, 130–133
geomembranes for, 134–136
geosynthetic clay, 136–138
postclosure monitoring, 138
vertical barriers, 51–58
Containment technology:
background, 1–3
overview of applications, 3–4

Contaminant migration:
containment technology, 2
soil-bentonite slurry trench cutoff walls, 17–18
Contaminant source zone, overview, 3–4
Coring control method, permeation grouting, 57
Cost analysis, bottom barrier construction, 78–80
Cracking pattern tests, hydraulic conductivity compatibility, 44
Critical depth of waste, bottom barrier construction costs, 78–80

Deep soil mixing, vertical barrier construction, 22–23
Desiccation cracking, bentonite-augmented soil caps, 134
Diaphragm (structural) cutoff walls, design and construction, 20–21
Differential settlement:
geosynthetic clay liners (GCLs), 124–125
surface control barriers (caps) construction, 142–143
Diffusive transport, mechanism of, 6–7
Dilatometer profiles, slurry wall state-of-stress measurements, 37–38
Directional drilling:
bottom barrier construction, grouting combined with, 67–70, 141
jet grouting, 65–67, 141
Direct shear testing, geosynthetic clay liners (GCLs), 122–124
Distortion definition, final cap barrier layer with, 106–107
DNAPL (dense aqueous phase liquids):
hydraulic conductivity compatibility studies, 47
remediation and containment technology, 1–3
Documentation, vertical barrier performance, 51
Double rod technique, jet grouting, 30
Downgradient vertical barrier, vertical barrier design, 12–14
Drainage layer for final caps:
materials for, 91, 95–100
geocomposite drain, 95, 99
geonet drain with geotextile filter, 95, 98
gravel drain with geotextile filter, 95, 97
sand drainage layer, 95–96
pipe capacity and slope, 95, 100
Ductility, bottom barrier construction, permeation grouting, 65

Envirowall composite cutoff wall, 23–24
Excavation techniques, soil-bentonite slurry wall, 16–17

Field stress conditions, slurry wall performance and, 35–40
Filter cake, soil-bentonite backfill, 50
Filter press testing, hydraulic conductivity compatibility, 44
Fines content, hydraulic conductivity and, 32
Floors. *See* Bottom barriers
Flow cone measurement, permeation grouting, 56
Foundation and/or gas collection layer, materials for, 110–115
geonet drainage material with geotextile filters, 110, 115
geosynthetic clay liners, 110, 116
geotextile drain, 110, 114
sand or gravel drain with soil filter, 110, 113
soil, 110–111
waste material, 110, 112
Freeze-thaw cycles:
geosynthetic clay liners (GCLs), 122
sidewall barrier construction, 141

Gas, cap control of, 129–130
Gas collection layer. *See* Foundation and/or gas collection layer, materials for
Geocomposite drainage materials, final cap drainage layer with, 95, 99
Geogrid materials, final cap surface layer with, 90
Geomembranes:
composite cutoff wall construction, 23–24
final cap barrier layer with:
compacted clay composite, 100, 104
geosynthetic clay composite, 100, 105
single liner, 95, 100, 102
geosynthetic clay liners (GCLs):
foundation or gas collection layer, 110, 116–117
shear strength and, 124
surface control barriers (caps), compacted soil composite with, 134–136
Geosynthetic clay liners (GCLs):
final cap barrier layer with:
engineering properties, 117–125
geomembrane composite, 100, 105
liner types, 110, 116–117
single liner, 95, 100, 103
foundation and/or gas collection layer with, 110, 116

Geosynthetic clay liners (GCLs) (*Continued*):
  surface control barriers (caps), construction quality control (CQC), 136–138
Geosynthetic erosion control material, final cap surface layer with, 84, 87, 90
Geotextiles:
  construction quality control (CQC), Claymax product, 137–138
  barrier layer, geosynthetic clay liners with, 110, 116–117
  drainage layer with:
    geonet drain with, 95, 98
    gravel drain with filter, 95, 97
  foundation and/or gas collection layer:
    drain with, 110, 114
    filter with geonet material, 110, 115
  surface layer with, 90
Gouy-Chapman theory, hydraulic conductivity compatibility, 43–44, 46–47
Grain size:
  grout type and, 27–28
  slurry wall performance, 32–33
  soil-bentonite backfill, 48–49
Granular activated carbon (GAC), contaminant transport, 7
Gravel:
  final cap drainage layer with, 95, 97
  foundation and/or gas collection layer with, 110, 113
Groundwater discharge and recharge zones, 9–10
Grout:
  cement-bentonite slurry trench cutoff walls, 18–19
  contaminant transport:
    placement during, geologic considerations, 8
    groundwater discharge/recharge zones, 9–10
  materials development for, 140
  vertical barrier wall construction with, 25–32
  vibrating beam cutoff walls, 21–22
  See also Chemical grout; Particulate grout, and specific grouting techniques
Grout cones, bottom barrier construction, 63–65
Grout curtains, vertical barrier walls, 15
Grouted bottom barriers:
  block displacement method (BDM), 70–72
  directional drilling, 67–70
  hydrofracturing and, 70
  injection grouting/sheet pile cutoff, 72–74
  jet grouting, 65–67
  materials for, 61–63
  permeation grouting, 63–65
GUNDSEAL:
  differential settlement measurements, 124–125
  geosynthetic clay liner hydraulic conductivity, 117–122
  shear strength, 122–124
GUNDWALL composite cutoff wall, 23–24

Hanging vertical barrier wall, 13–15
High-density polyethylene (HDPE):
  composite cutoff wall construction, 23–24
  geosynthetic clay liners (GCLs), 117, 124
Horizontal barrier installation, directional drilling, 69–70
Horizontal drilling, bottom barrier construction:
  jet grouting, 65–67
  permeation grouting, 63
Hydraulic conductivity:
  bentonite content, 33–34
  cement-bentonite slurry trench cutoff walls, 18–19
  consolidation pressure and, 35
  construction quality control (CQC) and, 5
  field stress conditions, 35–40
  fines content and, 32–33
  geosynthetic clay liners (GCLs), 117–122
    equivalency analysis, 120
  grouted bottom barriers, materials for, 62–63
  hydrofracturing and, 40–41
  permeation grouting, 26, 29–30
  test methods for, 41–42
  vertical barrier wall design and, 13–15
  water table fluctuations and, in soil-bentonite, 34–35
Hydraulic control, contaminant transport, 8–9
Hydrofracturing:
  bottom barrier construction, 70
  hydraulic conductivity, wall performance and, 40–41
  permeation grouting, 26
  See also Block displacement method (BDM)
Hydrogeologic conditions, contaminant transport, 8–10

Infiltrometer tests, construction quality control procedures, 5
Interim caps, characteristics of, 126

Jet grouting:
bottom barrier construction, 65–67, 141
directional drilling, 68–70
construction quality control (CQC) procedures, 57–58
vertical barrier wall construction, 30–31

Kaolinite, hydraulic conductivity compatibility, 43–44
Keying, soil-bentonite slurry trench cutoff walls, 17

Landfills, caps over, 129
Leachate:
contaminant transport and, 6
hydraulic conductivity compatibility, 43–48
natural bottom barriers for, 60–61
LNAPLs (light nonaqueous phase liquids), 4
vertical barrier design, 13–15

Marsh funnel, permeation grouting, 56–57
Methylene blue titration test, backfill quality control, 54
Microtunneling, injection grouting and, 76–77
Mix testing, cement-bentonite quality control, 55–56
Moisture content, compacted soil quality control, 130–131
Molecular diffusion, contaminant transport, 7

NAPL (nonaqueous phase liquids):
hydraulic conductivity compatibility studies, 47
soil-bentonite slurry trench cutoff walls, 18
Natural bottom barriers, characteristics of, 60–61

Optical survey techniques, compacted soil quality control, 132–133
Over consolidation ratio (OCR), slurry wall state-of-stress and, 36–38
Overlaps, geosynthetic clay liners (GCLs), 121–122, 125

Particle size distribution, compacted soil quality control, 130–131
Particulate grout, vertical barrier construction, 25
Paving material:
final cap barrier layer with, 107, 109–110
final cap surface layer with, 84, 89–91
Permeability of soil:
backfill quality control, 54
cement-bentonite:
quality control, 55
slurry trench cutoff walls, 18–19
compacted soil quality control, 131–132
grout type and, 27
grouted bottom barriers, materials for, 61–63, 141
natural bottom barriers, 60–61
Permeameter, hydraulic conductivity testing with, 41–42
Permeation grouting:
bottom barrier construction, 63–65
directional drilling, 68–70
construction quality control (CQC), 56–57
vertical barrier construction, 25–30
Piezometers:
permeation grouting, 57
conductivity, 29–30
jet grouting, 58
Plastic concrete slurry walls, design and construction, 19–20
Plume migration, contaminant transport, 8
Point injection technique:
directional drilling, bottom barrier construction, 68–70
permeation grouting, 26
Polymers, barrier materials from, 140. *See also* specific polymers
Portland Cement:
bottom barrier construction, 62–63
jet grouting, 30, 67
particulate grout with, 25
Postclosure monitoring, surface control caps quality control, 138
Precipitation:
assessment of overall cap containment, 126, 128
contaminant transport and, 6
Primary control, permeation grouting, 56–57
Proctor test, compacted soil quality control, 131
Protection layer for final caps, materials for, 91–94
''Pump and treatment'' methods, vertical barriers and, 11
Punctures, geosynthetic clay liners (GCLs), 121–122

Quality control:
containment technology, 4–5
hydraulic conductivity testing, 41–42
permeation grouting, 28–29

Radioactive waste, cobbles in final cap protection layer for, 91, 94
Radio tracking, bottom barrier construction, directional drilling, 69–70
RCRA regulations, surface control barriers (caps), geomembrane composites, 134–135
Reactive containment barriers, biotic/abiotic contaminant degradation, 7
Records of Decision (RODs), containment technology applications, 3–4
Rectangular microtubules, bottom barrier construction, side by side pushing, 77–78
Recycled/reused waste material, final cap protection layer with, 91, 93
Remedial measures (remediation), containment technology, 1–3
Residual shear strength, geosynthetic clay liners (GCLs), 123–124

Sand:
  final cap drainage layer with, 95–96
  foundation and/or gas collection layer with, 110, 113
Secondary control, permeation grouting, 57
Seepage barrier, defined, 11
Seismic considerations:
  bottom barrier construction, quality control (CQC), 81–82
  contaminant transport, 9
Self-sealing, geosynthetic clay liners, 117, 121–122
Shear strength, geosynthetic clay liners (GCLs), 122–124
Sheet pile cutoff, bottom barrier construction, injection grouting and, 72–74
Silty clay, natural bottom barriers with, 60–61
Single-rod technique, jet grouting, 30
Site dimensions, bottom barrier construction costs, 79–80
Sleeve pipe injection technique, permeation grouting, 26, 29
Slope angle of cap, assessment of overall containment, 126–128
Slump testing, backfill quality control, 54
Slurry walls:
  additive materials for, 140
  construction quality control (CQC):
    cement-bentonite slurry testing, 55
    soil-bentonite slurry testing, 52–53
  excavation techniques, 17
  performance factors, 32–41
    bentonite content, 33–34
    field stress conditions, 35–40
    grain size distribution, 32–33
    hydrofracturing and hydraulic conductivity, 40–41
    pressure consolidation, 35
    water table fluctuation, 33–35
  Soilsaw barrier wall, 31–32
  stratigraphic considerations, 8–9
  vertical barrier wall design, 15
  See also specific types of slurry trench walls, e.g. Soil-bentonite slurry trench cutoff wall
Soil:
  final cap protection layer with, 91–92
  foundation and/or gas collection layer with, 110–111
  See also Borrow soil; Compacted soil; Topsoil
Soil-bentonite slurry:
  backfill design, 48–51
  field stress conditions, wall performance and, 35–40
  grain size distribution and wall performance, 32–33
  hydraulic conductivity compatibility studies, 44–48
  trench cutoff walls, applications, 15
    construction quality control (CQC), 51–54
    design and construction, 16–18
Soilsaw barrier wall, vertical barrier construction, 31–32
Sorbents, diffusive contaminant transport, 7
Sorption processes, contaminant transport and, 7
Specific gravity measurement, permeation grouting, 56–57
State-of-stress:
  sidewall barrier construction, 141
  slurry wall performance and, 36–40
Steel reinforcement, diaphragm (structural) cutoff walls, 20–21
Steel sheet pile cutoff walls, 24–25
Steerable drill head, directional drilling, bottom barrier construction, 67–70
Stratigraphic conditions, contaminant transport, 8–9
Surface control barriers (caps):
  construction quality control (CQC), 130–138

bentonite-augmented compacted soil, 134
compacted soil materials, 130–133
geomembrane materials, 134–136
geosynthetic clay materials, 136–138
postclosure monitoring, 138
final cap components, 84–110
barrier layer materials, 95, 100–110
drainage layer materials, 91, 95–100
foundation and/or gas collection layer, 110–116
protection layer materials, 91–94
surface layer materials, 84–91
future construction techniques, 142–143
geosynthetic clay liners (GCLs) for barrier layer, 110–125
interim caps, 126
overall containment assessment, 126–130
overview, 83–84
test cap, proposed construction of, 143
Surface layer of final cap, alternatives for, 84–91
cobbles, 84, 88
geosynthetic erosion control material over topsoil, 84, 87
paving material, 84, 89
topsoil layer, 84, 86
Sword excavator, Zublin barrier system, 74–76

Tensile strain, geosynthetic clay liners (GCLs), 124–125
Tiltmeters:
bottom barrier construction, 81–82
hydrofracturing, 70
Tilt table testing, geosynthetic clay liners (GCLs), 122–124
Topsoil:
final cap barrier layer with, 106
final cap surface layer with, 84, 86, 90
Transport of contaminants:
geologic and hydrogeologic considerations, 8–10
groundwater recharge and discharge zones, 9–10
seismic considerations, 9
stratigraphic considerations, 8–9
mechanisms for, 6–7
Trench testing:
construction quality control (CQC), 53
soil-bentonite slurry wall design and, 16–17, 55
Triple rod technique, jet grouting, 30, 67
Tube-a-manchette technique:
directional drilling, bottom barrier construction, 68–70
permeation grouting, 26, 29

Upgradient barrier, vertical barrier design, 12–13

Vault containment, sheet pile cutoff and grouting for, 72–74
Vertical barriers:
applications, 11–14
circumferential design, 11–12
compatibility, 42–48
construction quality control (CQC), 51–58
cement-bentonite cutoff walls, 54–56
jet grouting, 57–58
permeation grouting, 56–57
soil-bentonite cutoff walls, 51–54
overview, 11
performance documentation, 51
sidewall barriers, 141
slurry wall performance factors, 32–41
bentonite content, 33
consolidating pressure, 35
field stress conditions, 35–40
fluctuating water table, 33–35
grain size distribution, 32–33
hydraulic fracturing effects on conductivity, 40–41
soil-bentonite backfill design, 48–51
test methods, 41–42
types of, 14–32
cement-bentonite slurry trench cutoff walls, 18–19
composite cutoff walls, 23–24
deep soil mixing, 22–23
diaphragm (structural) cutoff walls, 20–21
grouted barriers, 25–32
plastic concrete cutoff walls, 19–20
soil-bentonite slurry trench cutoff walls, 16–18
steel sheet pile cutoff walls, 24–25
vibrating beam cutoff walls, 21–22
See also specific barrier designs, e.g. Circumferential barriers
Vertical drilling, bottom barrier construction, jet grouting, 65–67
Very low density polyethylene (VLDPE), geosynthetic clay liners with, 117
Vibrating beam cutoff walls, 21–22

Waste material:
  compressibility of, in underlying landfills, 129
  foundation and/or gas collection layer with, 110, 112
  final cap barrier layer with, 107–108
  final cap protection layer with, 91, 93
Water/cement ratio, permeation grouting, 56–57
Water quality, slurry wall construction, 52
Water table, fluctuations in, slurry wall performance and, 33–34
Weathering mechanisms, compacted soil quality control, 133
Wet-dry cycles:
  geosynthetic clay liners (GCLs), 121–122
  surface control barriers (caps) construction, 142

Zublin barrier system, 74–77, 141